Creating High-Quality Vegetation for Games

This is a start-to-finish guide on how to build high-quality vegetation for video games. This book gives readers the fundamentals of the craft and the skills needed to confidently create natural environments.

Chapters cover everything from planning and gathering references, creating natural-looking ground cover with lots of variety, and building trees procedurally, to setting up a realistic wind shader before combining all assets into a scene in Unreal Engine 5.

This book will be of interest to all beginner and aspiring environment and vegetation artists looking to learn how to create high-quality vegetation, as well as more experienced artists looking to hone their craft.

Chico Spans is a Senior Vegetation Artist with multiple years of experience working at PUBG Corporation, Playground Games, Ubisoft Massive Entertainment and Epic Games (Quixel), as well as a Mentor at Vertex School.

Creating High-Quality Vegetation for Games

Chico Spans

CRC Press
Taylor & Francis Group
Boca Raton London New York

CRC Press is an imprint of the
Taylor & Francis Group, an **informa** business

Designed cover image: Chico Spans

First edition published 2025
by CRC Press
2385 NW Executive Center Drive, Suite 320, Boca Raton FL 33431

and by CRC Press
4 Park Square, Milton Park, Abingdon, Oxon, OX14 4RN

CRC Press is an imprint of Taylor & Francis Group, LLC

© 2025 Chico Spans

Library of Congress Cataloging-in-Publication Data
Names: Spans, Chico, author.
Title: Creating high-quality vegetation for games / Chico Spans.
Description: First edition. | Boca Raton, FL : CRC Press, 2025. | Includes bibliographical references and index. | Summary: "This is a start-to-finish guide on how to build high-quality vegetation for video games. The book gives readers the fundamentals of the craft and the skills needed to confidently create natural environments. Chapters cover everything from planning and gathering references, creating natural-looking ground cover with lots of variety, and building trees procedurally, as well as how to set up a realistic wind shader before combining all assets into a scene in Unreal Engine 5. This book will be of interest to all beginner and aspiring environment, and vegetation artists looking to learn how to create high-quality vegetation, as well as more experienced artists looking to hone their craft"— Provided by publisher.
Identifiers: LCCN 2024053396 (print) | LCCN 2024053397 (ebook) | ISBN 9781032794808 (hbk) | ISBN 9781032785646 (pbk) | ISBN 9781003492283 (ebk)
Subjects: LCSH: Video games—Programming. | Plants—Computer simulation. | Digital video.
Classification: LCC QA76.76.V54 S63 2025 (print) | LCC QA76.76.V54 (ebook) | DDC 794.8/151—dc23/eng/20241205
LC record available at https://lccn.loc.gov/2024053396
LC ebook record available at https://lccn.loc.gov/2024053397

ISBN: 9781032794808 (hbk)
ISBN: 9781032785646 (pbk)
ISBN: 9781003492283 (ebk)

DOI: 10.1201/9781003492283

Typeset in Times
by codeMantra

Access the Support Material: www.routledge.com/9781032785646

Contents

Acknowledgments

This book would not be where it is without the help of the listed people; they have all spent time and effort reading, reviewing, and suggesting improvements.

Writing a book takes effort and endurance. Having people show interest, review, and comment on the content is an incredible motivator to keep going, and each one of them did so unconditionally. For this, they have my gratitude.

Daniel Stok, whom I have spoken to almost daily, has given me advice and support and reviewed the chapters and figures. Thank you for always making yourself available to help out.

Sigge Sandström, for going above and beyond reading and reviewing the chapters and catching errors with an incredible eye for detail.

Ivan Stanić, thank you for teaching me some new things about SpeedTree, helping clarify a bunch of figures, and taking the effort to read and review chapters.

Kyle Reece, thank you for reading and reviewing all chapters, asking questions about the content, and suggesting improvements. Your perspective has helped me make decisions and guided the way I try to explain things.

Daehan Adams, thank you for reading and reviewing. Your ideas have helped improve and clarify every chapter.

Péter Foltán, for reading the chapters and letting me know what you think.

Tina Lyu, for your feedback on some of the images.

Josefine Holmberg for supporting me mentally while writing the book and helping me free up time to do so. The shoulder rubs, coffees, and chats have all helped me push forward.

Freya, thank you for keeping my feet warm and keeping me company throughout. I'm unsure if it was for me or the treat jar on the desk, but it was appreciated, nonetheless.

Lastly, thank you, for purchasing this book and trusting me to teach you something new.

About the Book and Author

ABOUT THE BOOK

The book's goal is to set you up for long-term success and to teach you to be comfortable building vegetation by hand and control the software you use rather than settling for the results presented by a tool.

The focus is on teaching you the process rather than what buttons to press, ensuring the information gained will serve you for the rest of your career rather than the next major software update. That said, the book will clearly describe which steps are taken and how to achieve them, so if you are a beginner, this book can still be followed along.

The book will teach you to become an independent Vegetation Artist with a solid understanding of the artistic and technical aspects of creating vegetation. You will learn how to build photorealistic content without relying on online libraries using tools accessible to everyone.

Within Vegetation Art, there are two main routes: hand-crafted and procedural; the latter is, in many cases, quicker when a lot of variety is required but does not teach transferable skills as efficiently. For that reason, the book will explore both with a heavier focus on handcrafting, ensuring the reader understands what to look for and how to deconstruct references into logical elements. The same thought process can then be applied to the procedural aspect, which we will explore in the book's latter half.

When procedural tools are used, the book explains them so that the reader remains in control and understands why specific settings are changed.

The book is aimed at people who would like to explore vegetation across all levels; the information inside will be relevant for beginners setting their first steps in the craft but will also speak to industry professionals who wish to expand their knowledge on vegetation.

To follow along with the book, you will require the source files, and these can be found here:

www.routledge.com/9781032785646

ABOUT THE AUTHOR

Chico Spans embarked on his journey into the industry in 2010, dedicating four years to his studies at Grafisch Lyceum Utrecht. After this education, he worked at Sticky Studios for a year, contributing to projects for renowned clients such as Warner Bros. and Disney.

His educational journey progressed at Breda University of Applied Sciences, where he seamlessly blended academic endeavors with freelance vegetation work on *PlayerUnknown's Battlegrounds (PUBG).* This endeavor proved fruitful, as the game soared to become one of the best-selling PC games, generating over $14 billion in revenue and selling 75 million units in software sales.

Expanding his horizons, Chico relocated to the UK and joined Playground Games to work on *Forza Horizon 4.* The title earned critical acclaim, winning a BAFTA award and a stellar score of 92 on Metacritic.

A new chapter unfolded as Chico pursued a career as a Vegetation Artist at Ubisoft Massive, where he worked on the highly anticipated *Avatar: Frontiers of Pandora,* which, upon release, was seen as one of the best-looking games of the year.

At the time of writing this book, Chico is employed at Epic Games, which is commonly known for *Fortnite* and creating the Unreal Engine. Within the company, Chico is actively engaged in various internal initiatives, pushing the boundaries of technology and elevating the quality of photorealistic plants and trees.

Beyond his professional pursuits, Chico is a Mentor at Vertex School, providing guidance and coaching to diverse students. He has also developed a comprehensive Vegetation course, available on the school's website.

Chico's extensive experience in the industry, spanning nearly a decade and with a specific focus on vegetation art for the past six years, positions him as a qualified author for this book.

As a frontline innovator, he continually pioneers new techniques aligned with the latest technology. Drawing from personal and professional experiences, Chico brings a wealth of knowledge to guide and mentor individuals within the industry.

Introduction

1

Making vegetation is complex, and the fact that you are reading this book means you have either already figured that out or, more excitingly, are about to.

So why is it that vegetation is challenging to build? One of the main factors is that if we look at how games are built, many aspects have developed over time, but how we approach vegetation remains essentially unchanged. In simplified terms, we are putting flat geometry with a branch texture on top of tubes; if you look at screenshots from games such as *The Legend of Zelda: Ocarina of Time* or *Halo: Combat Evolved*, you will notice that that is how it was done. For your reference, both these games are over twenty years old by now.

This is because vegetation consists of complex hierarchical structures, leaves attached to stems, stems attached to branches, and branches attached to trunks; if we built this out, it would be too expensive to run in real-time. For this reason, we build a cheaper abstract of the real thing when building vegetation.

To build this cheaper abstract, we need to use many tricks that require an in-depth understanding of modeling tools and game engines on top of complex scene management, and we will quickly spiral down into a whirlpool of technical considerations once we start adding destruction, wind, vertex colors and level of detail models (LODs) into the mix.

With that being said, innovations are lurking on the horizon; with the release of Unreal Engine's Nanite technology, we are starting to rethink our approach to building vegetation; the book will shine a light on both traditional methods as well as exploring what the future might bring.

Another reason building vegetation is hard is that everyone in the world is familiar with how it looks, meaning that if it is not built correctly or convincingly, it can quickly start looking off, and we do not want that; we want our vegetation to look great!

If you are still here after reading all of that and feeling ready to tackle the many technical and artistic hurdles ahead, congratulations! You have just set the first step in becoming a Vegetation Artist, and we are about to set off on an exciting and informative journey on how to overcome these obstacles.

I hope you enjoy the book, and I am looking forward to seeing the result.

Planning and Gathering References

2

IMPORTANCE OF REFERENCES

One of the most critical steps is gathering and organizing your references; if you are following this book and have the possibility, I highly recommend going out and taking many pictures of the subject you are trying to recreate and applying the techniques shown in the book to your vegetation. If you are unable to do that, you can find my reference file here:

...\CreatingHighQualityVegetation\ref\CRC_RefBoardCreatingVegetation.pur

One of my teachers once said, *"Your work will never be better than your best reference."* And I could not agree more. When collecting references, there are a couple of things to remember, but most importantly, make sure you do it deliberately. If you are starting with a project, it is okay to have many images exploring different topics, biomes, color schemes, etc., but at some point, you must cull down your selection and zoom in on more specific areas.

Another thing I would like you to keep in mind is that references are not just for you to get started but will be a fallback point throughout the project if you need to find some inspiration or ideas on where to take it next. See it as the foundation and make it as strong as possible.

One of the best tools out there for organizing and maintaining references in my opinion is PureRef; the tool can be downloaded at https://www.pureref.com/ for free – but if you can, I highly recommend donating to support the tool in the future.

When you open up PureRef, you will be greeted with an empty canvas that overlays on top of your other windows, and you can move this canvas around by Right-Clicking on it and moving your mouse around, to see the controls you can Right-Click in the canvas and select Help or use the shortcut Ctrl+H

Take some time to get familiar with the tool before moving onto the next page.

DOI: 10.1201/9781003492283-2

NAVIGATING PUREREF

PureRef has a very simplistic layout, which can be confusing if you have just opened the tool, but after spending some time with PureRef, this quickly becomes second nature.

To get images into PureRef, you can do a couple of things:

- You can drag one or multiple files into the canvas.
- You can Copy and Paste images from online sources and, in some cases, even drag the image from the webpage into the pure ref canvas.

Once you have your images in PureRef, I would first like to Normalize all the images so they roughly get the same scale. To do this, select all your images, Right-Click, go to Images, Normalize, select Scale, or use the shortcut Ctrl+Alt+Down.

After Normalizing, I select similar images and press Ctrl + P to arrange them with an optimal layout.

You will notice that if you perform an action with your mouse instead, the shortcut is noted next to the action performed; it Is recommended that you familiarize yourself with shortcuts as they will make your workflow more efficient.

SELECTING GOOD REFERENCES

Once you have all your images in PureRef, you should get a good overview of what you are looking at, and it is time to do some quality assurance. When collecting and looking at references, there are a couple of things you should keep in consideration.

Use Real-Life References

Vegetation, even when stylized grows by a set of rules and the best way to find these rules is by looking at references taken from the real world – if you have the possibility, I would suggest going out and looking at the subject in person, taking photos of points of interest and expanding your collecting of references, if that is not an option usually you can find really good references if you know the common and Latin name of the asset you are trying to build. Googling the Latin name will usually give you more scientifically correct references, which in many cases work great for recreating the asset digitally.

The other benefit of using real-life references versus paintings, games, or artistic mediums is that you get it from the source; when looking at games for reference, you risk copying mistakes made by another artist and losing the opportunity to make your own artistic choices.

References for Accuracy and References for Inspiration

This contradicts the previous point slightly, but splitting up your references into real-life references and images for inspiration is possible. Say you see a cool scene in a movie that inspires you; you can have that in your reference board for inspiration on mood, composition, or color scheme, but it needs to be supported by real-life references that tell you about the individual elements and how they are constructed.

Select Main References

Once you have done the initial culling and are happy with your selection, picking a couple of main references is important. These images hold the essence of what you are trying to recreate, organized from large to small. Starting with two or three images that contain the whole scene supported by one or two main references for each element within this scene, look at the reference file provided with the book for a good example of this structure (Figure 2.1).

FIGURE 2.1   Main references and main elements.

Organize from Large to Small

Now that the references are organized into main references and their main elements, we can organize them further, creating clarity and ensuring we are looking at the right thing.

As an example, let's have a look at the grass references. The *Overview References* show how the grass looks in the context of a scene; this is the end result, but it does not tell you what to build. Making this more granular when looking at the *Close up references*, where we can start seeing a little bit about how the grass grows, how tall it is, how the blades bend and twist, and what the ratio between healthy and decayed leaves is. Finally, *Grass Blade References* show us each leaf and how they are connected to the stem – in addition, these references will serve as a great starting point when building the grass in a later chapter (Figure 2.2).

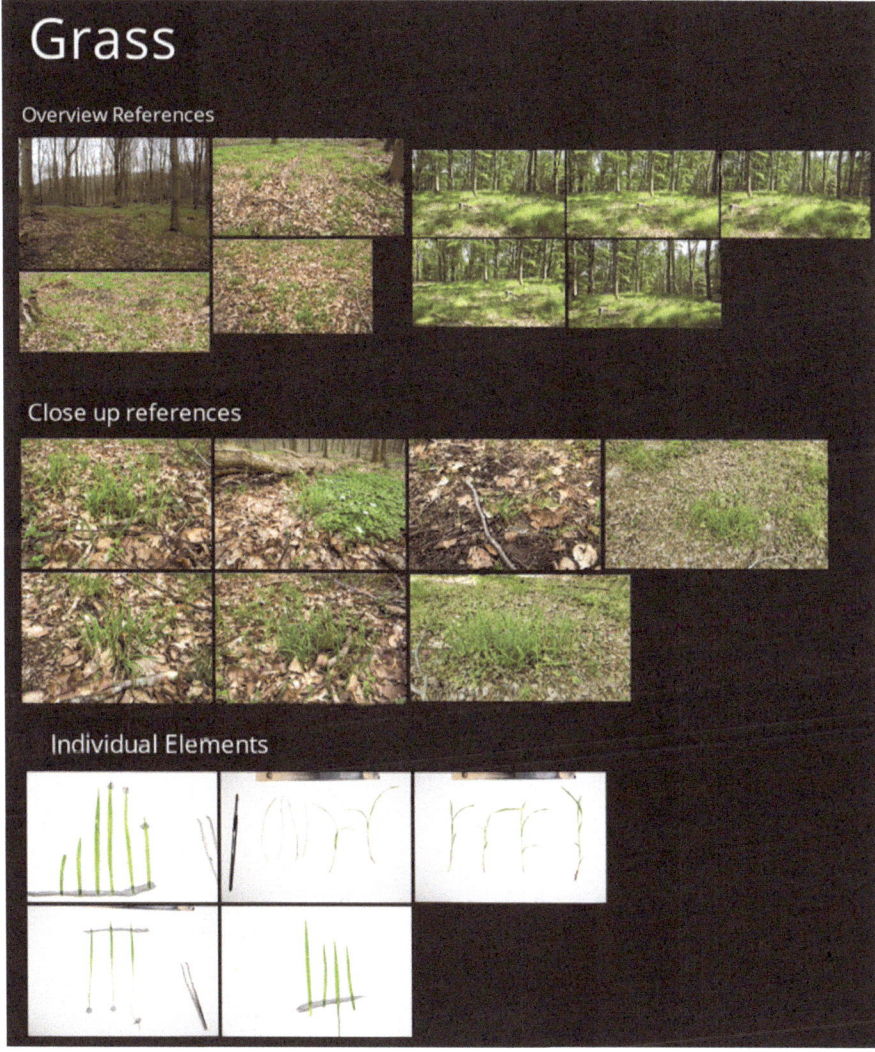

FIGURE 2.2 Grass references are an example of how to break down references into more granular elements.

Organize Structural Images Where You Can See the Way Things Are Constructed

One major advantage you can give yourself when building vegetation is understanding the ruleset and conditions in which a plant has grown. Every plant or tree is highly efficient and grows by a set of rules. These rules can be reverse-engineered, and understanding them will immediately make your vegetation look more convincing; I like to ensure that my references tell me at least the following things.

Phyllotaxy

The arrangement and number of leaves on the stem of a plant are genetically determined and are often a characteristic of a particular species; phyllotaxy is the term used to describe this arrangement of the leaves.

There are five main forms of phyllotaxy:

Alternate

If a plant has an alternate phyllotaxy, the leaves are found on nodes on both sides of the stem and move up in staged levels alternating between sides.

Opposite

With opposite phyllotaxy, the leaves also grow on each side of the stem, except they are found opposite.

Decussate

Decussate growing leaves are like opposite phyllotaxy growing on opposite sides of the stem, except each pair is found at a right angle to the previous pair.

Whorled

A plant with a whorled phyllotaxy means more than two leaves grow at each node.

Spiral

Spiral phyllotaxy is like alternate phyllotaxy, except they turn around the stem like a spiral staircase (Figure 2.3).

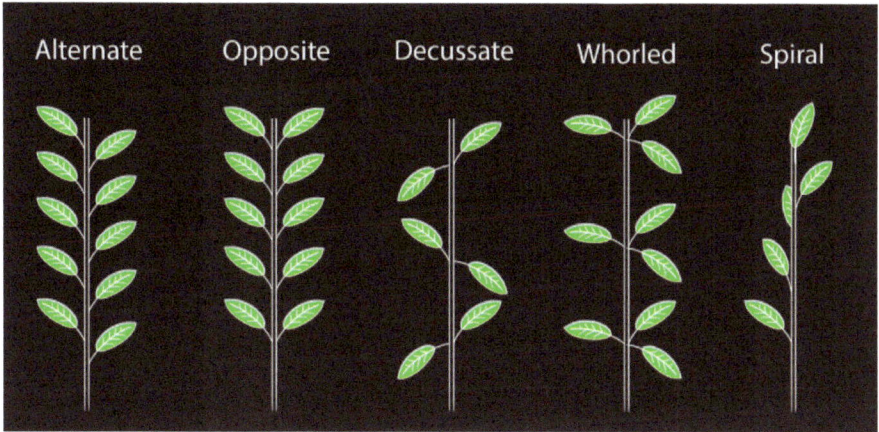

FIGURE 2.3 Different types of phyllotaxy.

I suggest that the next time you are outside, you look around to see if you can spot the phyllotaxy of nearby plants and trees. A solid understanding of this concept will help you throughout the rest of your career. And if nothing else, it can be a fun challenge while you're waiting on the bus.

Sunlight

Sunlight plays a huge role in where and how vegetation grows; a tree will try to find the quickest and most efficient way to grow toward the light each growth season. Therefore, if you look at a tree in a forest, they are usually very tall, with the canopy starting high up in the tree with many vertical branches trying to reach the sun, but on a field or a park, they will almost always have a hemisphere-like canopy, where the bottom branches are long and horizontal.

Identify what type of references you are using and ensure they match the topic you are trying to recreate; placing a hemisphere-like canopy in the middle of a forest will most likely not look natural (Figure 2.4).

Health and Growth Conditions

Another major factor that needs to be taken into account when dissecting references and planning ahead is health and growth conditions. A plant or tree grown in a greenhouse will look vastly different from the same species in a harsh desert; besides sunlight, the nutrients that are found in the ground will partially decide how the plant will grow. If there are not enough nutrients, the plant will for example have trouble growing longer spurts each growth season, growing very slowly, searching for light or water could result in very gnarly looking trees and plants, whereas rich ground usually results in longer and full vegetation.

FIGURE 2.4 A beech in a forest (left) and a tree without any competition (right).

These conditions will also decide the leaf color and saturation, leaf size, leaf and fruit amount, and whether the plant will bloom; make sure to get a rough idea of the conditions your subject is in before starting to build your block out.

Branch Transitions

Branch transitions are also a point of interest; if you can see your subject in real life, it would be good to take a couple of shots and see if there is a rule to how branches grow. Are they paired with leaves at the node? Do they come out straight or at an angle? Do they have a flare where they connect to their parent branch, or is it a straight transition? Small details like this are essential, and the book will explain how to build these accordingly in the next couple of chapters.

Leaf Clustering and Shape

At this stage, note how the leaves grow. Do they overlap? Is there more than one leaf per node, or does it have multiple? How do they bend? Are they convex or concave? Leaves can be unique per species or subspecies, so observing and imprinting them in memory is crucial to correctly identifying the species you are working on.

Unique Elements

Lastly, have a look at which unique elements you want to recreate, nature leaves its mark on trees, so look out for any broken branches, scars, or gashes and bulges – this is one of the only elements in the reference that is not likely to repeat across all your assets and will help give your asset a unique look and a bit more of a soul, these elements are important for a convincing and aesthetically pleasing result (Figure 2.5).

FIGURE 2.5 An example of branch transitions, leaf clustering, and unique elements.

This concluded the basics for gathering and analyzing references; I would like to emphasize the importance of this step again. Depending on the size of the project, it can take multiple hours to put a solid reference board together, and it is, in many cases, an evolving thing throughout the project. I advise you to spend some time browsing through your reference board to see if everything is covered, and it is a place you can return to get some inspiration.

Alternatively, if you are using the reference board provided, feel free to add some images of your own. This will help you shape an image of what you are trying to achieve personally and will help you tremendously throughout the book.

Now that we have covered references, it is time to build our first asset.

Creating
Ground Cover

3

INTRODUCTION

One of the most important things when building a forest scene is getting the ground cover right. Usually, there is a large variety of assets such as dead vegetation, saplings, weeds, and native species; striking a balance between all of these is essential in creating a believable forest floor; this chapter will cover the techniques to build most of those.

We will go over an initial introduction about creating groundcover, and we will then go over how to combine all pictures into a combined Photoshop file, also referred to as a texture atlas, before moving on to quick and efficient ways to remove the background.

After that, we will build all supporting maps, such as the Opacity, Roughness, and Normal Map, and explain what these are. Subsequently, we will move into Blender to create the geometry for the Grass and Wood Anemone. These meshes will then be used to rebake our initial atlas into a neatly optimized new atlas, which we will use to create our final grass clusters, including LODs and a Wood Anemone, you will then have the option to built the remainder of the plants: the Buttercup and Creeping Charlie, using the techniques you learned in this chapter.

In our source folder here ...*CreatingHighQualityVegetation\workfiles\source\source_undergrowth*, you will find pictures taken from multiple plants. This chapter will cover the grass and Wood Anenome, but the techniques explained are identical for the other two species.

Additionally, I will refer to our reference a lot in this chapter, so make sure you open up the ..*CreatingHighQualityVegetation\ref\RefBoardCreatingVegetation.pur*.

There are many excellent ways to capture data for smaller assets, such as leaves, but photometric scanning generates the best results. Photometric scanning uses a fixed camera that takes multiple images with different lighting conditions to estimate a normal direction. Additionally, you can use a polarizing filter to estimate a roughness value.

The drawback of this technique is that it is not accessible to everyone. Therefore, this chapter will focus on creating the necessary data by manipulating the Albedo extracted from a single photo taken from above.

DOI: 10.1201/9781003492283-3

My setup consisted of a white A3 paper to make sure I have a good contrast with the background, this will help remove the background more easily. A DSLR camera on a tripod and a daylight-adjusted lamp, but you can achieve similar results using a phone camera, the caveat being that you will need to be able to shoot your photos in a RAW format, else the phone camera will enhance the images. This is unwanted when generating textures from photo data.

If you want to create a different type of vegetation than described in the book, you are welcome to use your own photos, but if you do not have access to the vegetation you want to create, you can use the pictures provided in the book or find something similar on Google Images.

COMBINING ALL PHOTOS IN AN INITIAL ATLAS

Out of the box, just the photos might not be that comfortable to work with in other software, so I will start with manipulating some of the data in Photoshop. So, open Photoshop and create a new file, set the dimensions to Pixels, and make it 8192 by 8192. This value will work best if you use source images that fill up that space, if you are using images that are substantially smaller than this, it is better to divide the resolution in half, so go for 4096 by 4096 or even 2048 by 2048. This will not be the resolution we end up using, but for the time being, this will give us a large canvas to work with while maintaining as much of the resolution from the sources as possible. It's a good idea to save the file and give it a location. I called mine PSD_GroundCover_InitialSelection. It can be found here: ...\CreatingHighQualityVegetation\workfiles\photoshop\PSD_GroundCover_InitialSelection.psd.

In this file, I will put all my photos taken to create textures and group them up by their asset name; you can find these photos in the source folder mentioned in the introduction section of this chapter. If you get stuck or are otherwise unable to complete the steps explained in this chapter, all the steps will be separated in this file, and you can use it to follow along with the rest of the book.

Even though the rest of the chapter will focus on the Grass and the Wood Anemone, I will include the Buttercup and Creeping Charlie. Packing all of these in a single texture will save memory and be more efficient for Unreal Engine later.

Your file should look somewhat like Figure 3.1.

The first thing I like to do is cut out any excess material, so hit M or select the Rectangular Marquee Tool and go through all the images, picking just the leaves and stems; use Ctrl+Shift+I to invert your selection and hit Delete; this will get rid of the excess parts of the layer. Using Ctrl+D after to get rid of the Marquee selection.

It could be that your photos are imported as Smart Objects; in that case, choose the images, Right-Click the layer, and click Rasterize Layer, they will not be editable otherwise.

FIGURE 3.1   A photoshop file with all four source photos loaded in inside groups named after the plant species.

REMOVING THE BACKGROUND

Now that we have eliminated the excess background, it is easier to arrange the images next to each other and fix them up further. The goal is to eliminate the backgrounds, and there are multiple ways of achieving this. For the examples, I will use the Grass.

- Start by laying out the grass next to each other. You will notice I used some gray putty to keep the grass from folding and curling away while I was taking the photos. This makes it slightly more challenging to remove the background in one go, so we will need to do some manual edits as well.
- Once laid out next to each other, we can select all layers and hit Ctrl+E to merge them. If you haven't already, I advise creating a black background layer to see what we are removing more easily. If you still have the Background layer, select it and hit Ctrl+I to invert the color, or make a new layer at the bottom of the layer stack and fill this with black.

The easiest way to remove white backgrounds is to double-click the Layer, which will open up the Layer Style menu. Then using the "Blend If" options at the bottom, you can move the slider on the right toward the middle. I set mine to a value of 192, then hold Alt to split up the arrow; this will create a range of when it is supposed to blend; in my case, 179 did the trick; do note that these values can differ depending on the image so play around with it until you get a similar result as Figure 3.2.

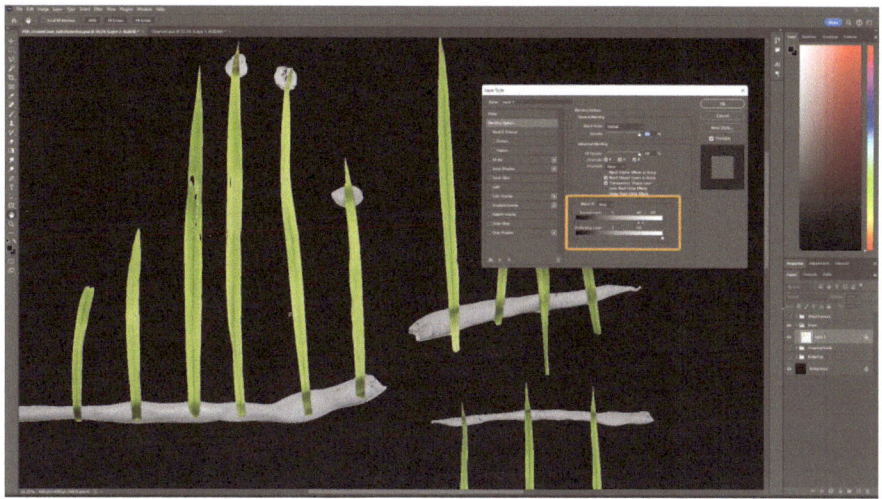

FIGURE 3.2   A photoshop file with the layer style window open, showcasing where the blend if settings are.

It is better to be aggressive to ensure we get rid of all the white outlines. You will notice that it does remove parts of the grass as well. For now, it is okay to see some gaps in the grass blades; in the following steps, we will explore a technique to restore this.

To remove any areas not affected by the blend layer, we will need to go in manually, so select the Grass layer and create a Layer Mask; in a Layer Mask, anything you paint black will become invisible. I used a pen tablet and a regular round brush as well as a combination of the Polygonal and Magnetic lasso to select the gray putty and mask it using the Layer mask. My advice would be not to overthink this too much; it's okay to have some small white parts left; further editing will most likely get rid of those, and there will be opportunities down the pipeline to fix errors made in this step.

If you have also used the Blend If function aggressively and are seeing gaps in your grass blades, it is important to get that back. To do so, you can duplicate the layer, add a completely black layer mask, making it completely invisible, remove the Blend If, make sure this layer is underneath the masked layer, and use a white brush to paint back the parts that you want to keep. Refer to the blue and orange marked areas in Figure 3.3 to see the layer stack and what it is affecting.

Lastly, we must remove the discoloration from the gray putty showing through the leaf. We can merge our previous layers but keep a copy of the original to ensure you can return and make any necessary edits. Select both layers, use Ctrl+J, then Ctrl+E to merge. Use the buttons at the bottom of the layer stack to create a Brightness and Contrast adjustment layer, hold Alt, and click on the line between the layers; this will ensure the adjustment layer only affects the layer underneath it. Then, use a soft brush with a low opacity, like ten percent, and slowly paint away the affected areas. Refer to Figure 3.3 to see the finished result marked in green.

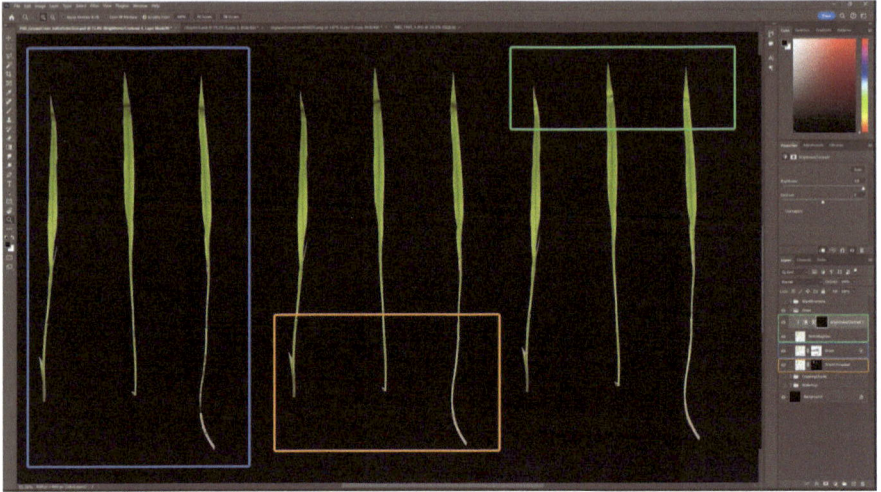

FIGURE 3.3 An overview of what happened in what layer and what the current photoshop file should look like.

You will notice that the Wood Anemone is harder to use the Blend If method on, as the leaves are white and, therefore, too similar to the white background. For those, I used the Magnetic Lasso and a manually painted mask.

After you have successfully removed all backgrounds from the images, I would like to organize each piece in a neater layout. To do this, I make a copy of all the layers and merge them together using Ctrl+E, then use the Polygonal Lasso, shortcut L to select the pieces I want to move and use Ctrl+J to make a new layer based on my selection. Do this for all pieces, and then use the Move Tool (V). In the top left corner, you can turn on Auto Select; this way, you can select the pieces you want to move and move them to the desired location. I like to keep the pieces that belong together somewhat in the same area. Refer to Figure 3.4 to see a before and after.

Once you have it organized, it is a good idea to do a final check on all your pieces to see if you have missed painting away any unwanted elements. You can do this by adding a mask to the root folder and masking away some last bits that you missed in previous steps.

It is beneficial to straighten out some of these shapes; this will allow us more freedom to deform them as we see fit once we start creating meshes for some of them. To straighten them, we will use the Puppet Warp tool in Photoshop, I will explain the steps I have taken, but if you have never seen or heard of Puppet Warp, now is an excellent time to play around with it a bit, and get a feel for it, before jumping in and doing it on all your assets. Additionally, always make sure you duplicate previous steps if you need to go back and make a change; now is a good time to do so, as Puppet Warp is a destructive workflow. Alternatively, you can save a copy of your photoshop file as a backup that way you can always go back to a previous step.

A destructive workflow locks you out of any previously taken steps, while a non-destructive workflow preserves older data. In this industry, striving for a

FIGURE 3.4 A before-and-after for the organizing step. Auto Select is highlighted in the top left corner.

non-destructive workflow is always best, as feedback will inevitably come, and you want to have positioned yourself to respond to that efficiently.

Once again, copy the previous Organizing folder, and if you want, collapse the layers. This will make the puppet warping a bit quicker. With a layer selected, go to Edit and choose Puppet Warp. In Puppet Warp Mode, you can place pins in your image and move those pins around to straighten out, for example, the stems. I advise you to use as few pins as possible, and in case it is necessary use Puppet Warp on the object multiple times.

Refer to Figure 3.5 to see Puppet Warp in action. Repeat this for all elements bent in the original photos, including stems, leaves, and flower pistils.

Once the straightening is done, we can take our initial atlas to Substance Designer and derive additional texture maps from it there; if you are stuck or unable to complete the steps above, please use the source Photoshop files instead; they can be found here: *..\CreatingHighQualityVegetation\workfiles\photoshop\PSD_GroundCover_Initial Selection.psd.*

To use the atlas in Substance Designer, we will need to save it out, so hit Ctrl + Alt + S and save it at a location of your choosing, I saved mine here: *..\CreatingHighQuality Vegetation\workfiles\designer\import* and called it "T_GroundCover_InitialSelection."

I have saved it as a .PNG and would like to use the opportunity to briefly touch on file extensions; throughout the industry, there are three major file extensions that are commonly used .PNG, .JPG, and .TGA and they all have a specific use case. PNG and TGA are both lossless formats meaning no quality loss occurs when saving the file, the difference between these two is that PNG will bake the Alpha Channel (or opacity) inside the image and TGA will maintain is as a separate channel, in our case I chose PNG because Windows is able to preview these files more easily. Later in the book we will use .TGA for our final exported textures because .TGA is considered industry standard.

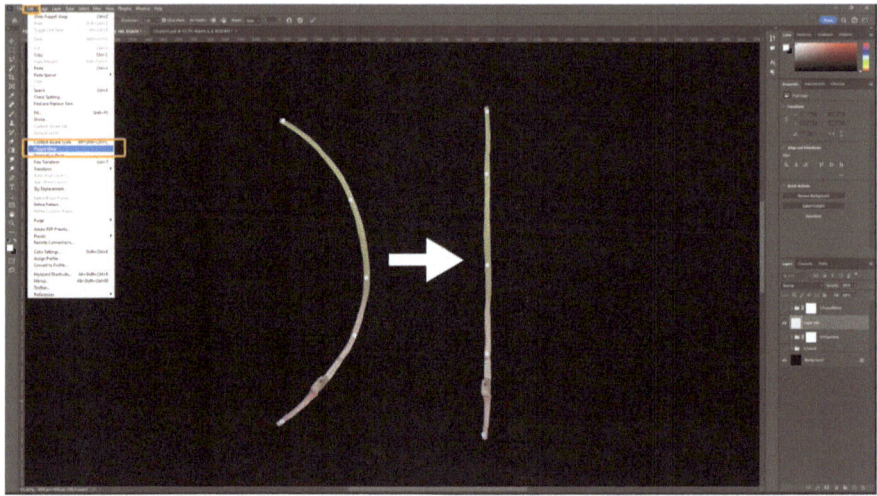

FIGURE 3.5 shows a stem being straightened using puppet warp and five pins; puppet warp is highlighted in the edit section.

Lastly, .JPG is mostly used when sharing images that are not intended for game assets or if the image size plays an important role, it is important to remember that when you save it as a .JPG the image gets compressed and loses quality, which is unwanted in many situations.

I like to divide my folder structure work files by the program I have used and I have an Import and Export folder where I place everything that goes in and comes out of that software.

Open up Substance Designer and make a new graph; I have called mine SBS_GroundCover_AtlasProcessing and saved it here: ..*CreatingHighQualityVegetation\\workfiles\\designer.*

If this is your first time using Substance Designer, here is a quick overview of the UI. It is an incredibly versatile and powerful tool, but an in-depth explanation is beyond this book's scope. So, if you are unfamiliar with it, I recommend spending a couple of hours clicking around, trying things out, and just having fun with it. This will make it significantly easier to follow along with the rest of the steps.

Refer to Figure 3.6 for a basic overview of the UI.

1. The explorer shows you which graphs you have loaded and which sub-graphs this contains; imported textures and meshes will also end up here.
2. The library shows all the embedded content, filters, and nodes inside Substance Designer, I hardly use this, so I tend to close this when editing the layout.
3. The graph view is where most of the work is done and where we will place down our nodes.
4. The 3D view shows you your texture under basic lighting conditions, it is possible to load in a custom mesh.

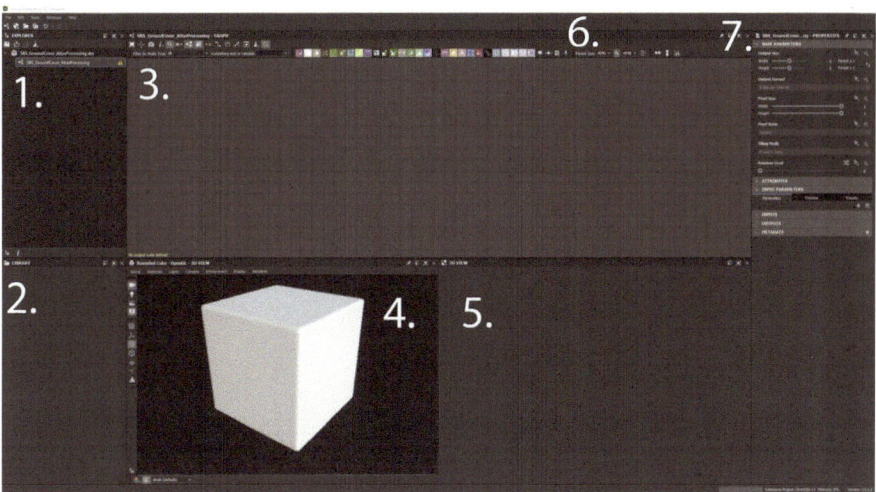

FIGURE 3.6   The UI of substance designer in its default state.

5. The 2D view shows your currently active node as it is being processed.
6. The parent size and toolbar, since our texture is 8192 pixels, we need to change that here.
7. Base parameters, this window will change depending on what node you have selected and is where we will set our values.

To import our texture, drag it into the viewport. You will be presented with two options, Link or Import if you select Link, the texture will update when the source file changes, if you choose to Import, it will be embedded into the substance file. I like to set my pipeline up in a way that updates things automatically, so I have opted to Link the resource. What this means is that when we now re export our texture from Photoshop, the updated version will automatically appear in Substance Designer.

The first and arguably one of the most important things we will build is an Opacity Map. An Opacity Map is a texture consisting of only black-and-white values, where White tells the engine what parts of the texture are opaque/visible and what will be transparent/invisible. This comes with some considerations.

When you use Opacity Maps and start layering meshes and objects with transparency on top of each other, this can become expensive quickly; this concept is described as Overdraw. The reason why this is expensive is that every consequent layer of opacity becomes more and more costly, and this can quickly escalate within vegetation. Therefore, when we get to the meshing step, we will try to reduce the areas that require an opacity map or eliminate it completely. Still, for our Substance graph, we will use the opacity maps to create a variety of masks that will help us generate additional texture maps from our Base Color.

For this part of the chapter, you can refer to the Substance file here ..*Creating HighQualityVegetation\workfiles\designer\SBS_GroundCover_AtlasProcessing.sbs.*

I highly recommend you have this file open while working along the chapter, as it will be easy to get lost in the explanation otherwise. Please note that the values used might not work precisely for your case, so feel free to experiment and play around with the numbers to get a result you think looks good.

For the explanation of the graph, refer to Figure 3.7.

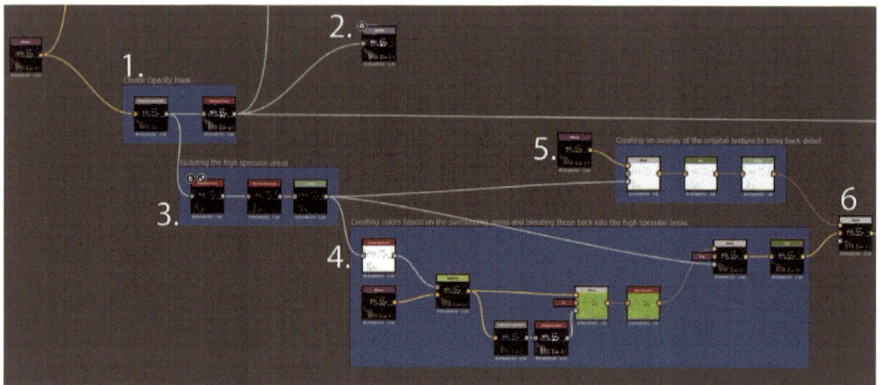

FIGURE 3.7   The first part of the graph describes steps taken to get to the final result.

BUILDING AN OPACITY MAP

1. To create an opacity mask, I used a Grayscale conversion node to turn my Base color into a grayscale map; this was plugged into a Histogram node with both Position and Contrast set to 1.
2. The result is put into an Output node. I set the Identifier to "O" and the Label to Opacity. Under the Usage panel, I added an item and set it to "opacity."
      You can now Right-Click in the graph and say "View Outputs in 3D View to show your outputs applied to the default mesh."
3. As you might have noticed, there is quite a bit of specular response left in the images. Specular response refers to the amount of light we see reflected in the images, this is unwanted as we want the texture to be as neutral as possible so we can do the lighting in engine later, any form of lighting information in the textures should be removed to the best of your ability, and we can use our grayscale texture and another Histogram Scan to isolate those areas. I used a Position Value of 0.44 and a contrast of 1, blurred this using a Blur HQ Grayscale node with a value of 0.1, and then used a levels node to bring in more blacks. The goal here is to get a mask that isolates the highlighted areas; if you want, you can put your base color in a Blend node, keep the settings default and plug the Levels node into the mask for the Blend; this way, you can preview the Blend node and see what areas are being isolated.

4. Inverting our previous result using an Invert Grayscale node, which is input into a Distance node as a mask together with our Linked Base color texture. What this does is remove any of the specular highlights and fill it with the distance node results; the distance node will take the pixels at the edges of the mask and stretch those pixels until it hits another pixel. At this point, most of the specular response should be gone, but the distance node generates quite harsh results; it could also be that you are left with some black values from holes in the texture. To remedy this, we take our Distance out and put it into a Grayscale conversion, run another Histogram Scan with the Position set to 1 to create a mask, and Blend the Distance node with a green color similar to the greens we have on the texture.

This is then slightly blurred with a value of 0.2 to remove the harsher lines created by the Distance node and blended once again with our Base Color texture and the Levels node created in Step 3; in my example, I had to up the Saturation a bit to match it with the surrounding elements, so I used an HSL node with the Saturation set to 0.52.

5. By removing the specular, we have lost some vein details, which we can bring back by using the levels node we made in step 3 as a mask and blending that with the original base color. We then turn that once again into a grayscale map and a levels node to decrease the contrast and slightly darken the texture, then blend that with Blending Mode set to Soft Light. The goal here is to bring some of the smaller-scale details back.

6. A Blend node blending the previous two steps, the Blend Mode is set to Soft Light.

You will see me mention the Blend Node a couple of times throughout the book. Whenever the Blend Type is not specified, it means I have left it in its default setting.

That concludes the first part of the Base Color graph, so we can move on to the second part, which is focused on bringing everything together. Refer to Figure 3.8.

1. The blend node from the previous steps.
2. We slightly blur the original Albedo to use it as a color overlay.
3. We convert our previous result into a Grayscale and use a levels node to again target any brighter values.
4. We blend in the blurred base color with an opacity of 0.2 using the Levels that target higher values as a mask; this way, we again slightly dampen the brighter values in our original Base Color.
5. As you might have noticed, all our previous actions have been affecting some parts of the texture negatively, here we use an SVG node to mask out the Buttercup and Creeping Charlie, and blend that with the imported Base Color to ensure the Grass and Wood Anemone stay as is. They did not have a big specular response when taking the photos and, therefore, required less work.

To create the SVG, make an SVG node, and say From New Resource, then double-click a node where the leaves are visible, and click on the SVG node only once, then use the Extrude Brush to brush over the parts you want to be affected.

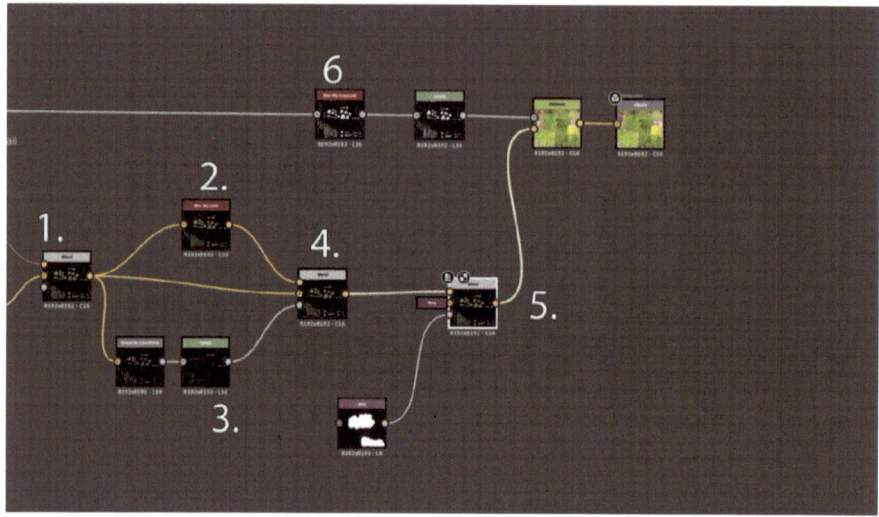

FIGURE 3.8   The second part of the base color graph.

Alternatively, you can paint these masks in photoshops and import them into Substance, I opt for SVG masks because it allows me to stay within Substance designers and reduce the amount of bridges I need to make between programs but feel free to make the masks whichever way is more accessible or quicker for you, there is no right or wrong here.

You will also notice that in the Blend mode, there is a "Bmp" node visible. This is a Docked node and contains the original imported texture. To Dock a node, you can select it and use shortcut D.

6. Our opacity mask goes into a blur node and then a level node to remove most of the gray values. The intention here is to Shrink our opacity mask a little and then plug the final Base Color into a Distance node. In this case, the Distance node is meant to create some texture padding.

Why Is Texture Padding Necessary?

When a texture is imported into an engine, the engine will perform all sorts of optimizations. One of those optimizations is "Texture Mipping" which means that over the distance, it will decrease the resolution of the texture. So a texture with a size of 2048 pixels might scale down all the way to thirty-two pixels over distance. When this happens on an Opacity Map, this will create inaccuracies and a mismatch in the Opacity and Base Color textures. If we leave the Basecolor background black, the black will start showing up in the mesh, so if we make the background a similar color, we prevent the mismatch from giving us any visually unpleasant results.

This wraps up the edits we need to make to the Albedo for the time being so that we can move on to the Roughness and Normal Map; these maps work closely together, so when working on these, I encourage you to experiment with different values and see what works best. It's best to do this in your target engine, but the Substance 3D Viewer will do the trick for a quick preview.

BUILDING A ROUGHNESS MAP

What Is a Roughness Map?

A roughness map is a value that dictates how rough or smooth a surface is; a roughness of 1 would indicate something like stone or coal, whereas a roughness of 0 would be a mirror. For vegetation, usually, we are within the 0.4–0.7 range.

With that in mind, let's set up an initial roughness map. Keep in mind that the values mentioned here are just suggestions. Your roughness value will depend slightly on personal taste, which reference you are going after, the strength of your normal map, and, for example, the weather conditions in your scene. If it is a rainy scenario, you probably want to go with a lower roughness than a desert scene.

We can keep this setup quite simple and add details once we start seeing some of our assets inside Unreal Engine.

Refer to Figure 3.9 to see my steps to build a roughness map.

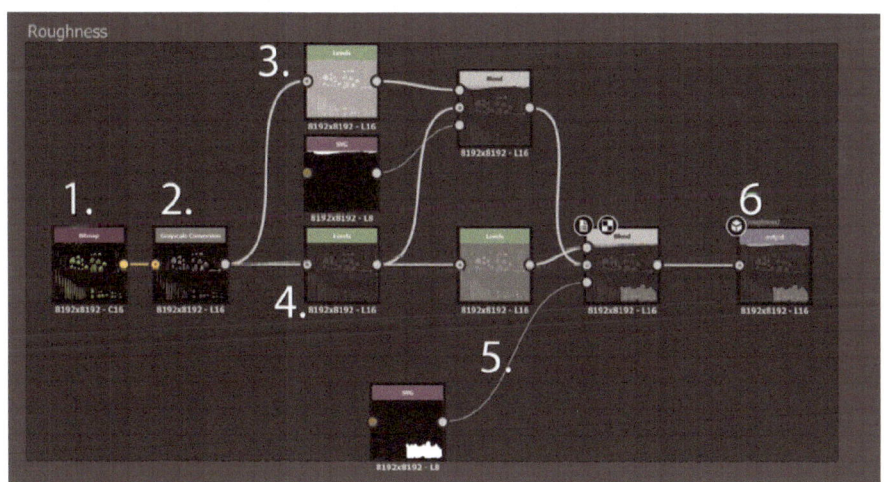

FIGURE 3.9 An overview of different steps taking to build a roughness maps.

1. Starting with our originally imported Albedo.
2. We use a Grayscale conversion node to make it cheaper for Substance to compile the nodes. A roughness map is always a single value, so we only need a grayscale map.

3. This levels node makes all the values brighter and rougher. This is then blended with our roughness values of step 4; the SVG node creates a mask, so we are only affecting the part of the atlas that contains the Wood Anemone.
4. Our base value for the roughness: in the end, this part only affects the buttercup atlas.
5. Another SVG mask and levels node, the levels node makes the base roughness only slightly brighter and therefore rougher; the SVG mask makes sure this is only happening for the creeping Charlie.
6. In An output node, the Label is set to Roughness and the Identifier to R inside the node settings; under components, add a component and make the Usage Roughness as well.

Once this is set up, you can Right-Click the graph and select "View outputs in 3D View" to preview your maps, compare the result to your references, and make sure you see a similar response; it could be a bit hard to tell at this point, as we do not have a normal map yet, so that is what we will look at next. I encourage you to double-check and adjust the roughness as you see fit when working on the Normal map.

Refer to Figure 3.10 to see my roughness in the 3D and 2D view.

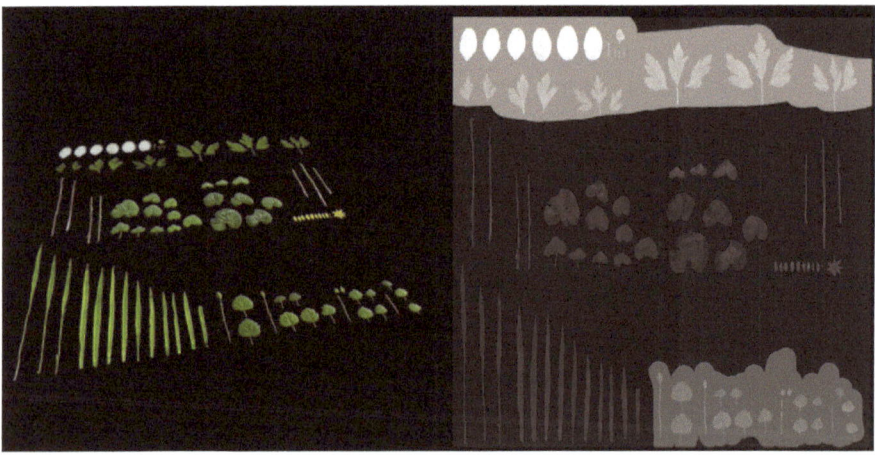

FIGURE 3.10   The 2D and 3D view of substance designer showcasing the current roughness map.

BUILDING A NORMAL MAP

With our roughness out of the way, we can move on to the normal map. We must bring in some extra painted maps from Photoshop or another tool for this texture. We will build different normal maps for our respective plants and then combine these into one normal map using Substance Designer.

So, What Is a Normal Map?

A normal map is a technique for faking the lighting on an otherwise planar mesh. It tells the Engine which way the light would bounce off and can, therefore, be used to generate details in a mesh without using extra polygons. The map uses different colors to define which direction the surface is going. Red defines the horizontal axis, Green the vertical axis, and blue refers to "depth"; this might be a bit abstract to understand initially, but if you work with them for a while, it will quickly become second nature.

After the chapter, I recommend you compare your preview with and without a normal map to get a good idea of what the texture brings to the table.

Ideally, a Normal Map is baked down from a high poly mesh or photometric scans, but in our case, when working with a single photo, it is important to layer up different maps normal maps with different details. This ensures you capture both the primary, secondary, and tertiary shapes.

We can do all of this within Substance Designer, you can use the source file found here ..\CRC\CreatingHighQualityVegetation\workfiles\designer\SBS_GroundCover_AtlasProcessing.sbs as a reference or follow along with the pictures and explanation.

I like to start with my primary shapes and build in more detail as I go along, so let's begin with the Stem and Leaf curvature. Both will use the Opacity map generated from the Albedo as a base.

1. An SVG node masking out the leaves of the Buttercup and Creeping Charlie, we mask this out because, as seen on the references, these two are concave while the others are convex.
2. The first input of the Blend node holds our SVG mask for the Buttercup and Creeping Charlie, and the second input holds an SVG node with all our stems marked. Those are combined into a mask for all the stems, and the Buttercup and Creeping Charlie leaves. This mask is used as a mask in step 5 to mask those away.
3. A blend node with the Opacity plugged into the first input, and the SVG stem mask into the masked input, ensuring we only edit the stems. Alternatively, a stem mask could be built in Photoshop and imported; this is then blurred with a value of 1.09 – I recommend you play around with these values to find a number that works for you. From there, we generate a normal using the Normal node and blur this slightly with a value of 0.1 to eliminate any stepping that occurs due to the low bit rate of the texture.
4. We repeat the same steps as in step 3, but this time using the masked leaves as an input, blurring those by 2.23, generating a Normal using the Normal node, and blurring this with 0.1 for the same stepping issue. They are then combined using the Normal Combine node.
5. Since the Wood Anemone and Grass are convex, we take our Opacity and invert it using the Invert Grayscale node. This way, we ensure the direction is down when generating a normal map later.

   The mask created in step 2 is used to blend in a white Uniform Color (docked using the D shortcut) using a Blend node. This leaves us with just the grass blades and anemone, which are then blurred by a value of 1.39 and converted to a Normal texture using the Normal node. We then once again blend both results using the Normal Combine node.

That concludes the Stem and Leaf Curvature. You can refer to the provided example file or Figure 3.11 to see what this result is expected to look like. I recommend checking both the 2D and 3D views to see your result to get a good idea of what it will look like in the engine later on.

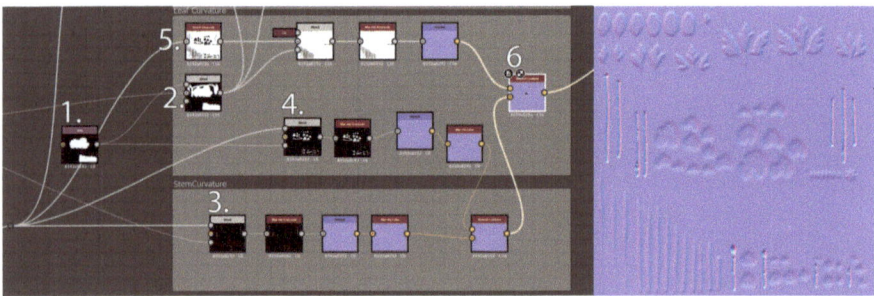

FIGURE 3.11 · The node graph for the stem curvature and leaf curvature part of the normal map, with the normal map visible on the right.

Referring to Figure 3.12, we can now have a look at the more detailed aspects of the normal texture. For this step, we will need to create a Vein Mask; refer to Figure 3.13 to see what mine looks like; alternatively, you can find the file here: ..*Creating HighQualityVegetation\workfiles\designer\import\T_GroundCover_VeinMask.png*.

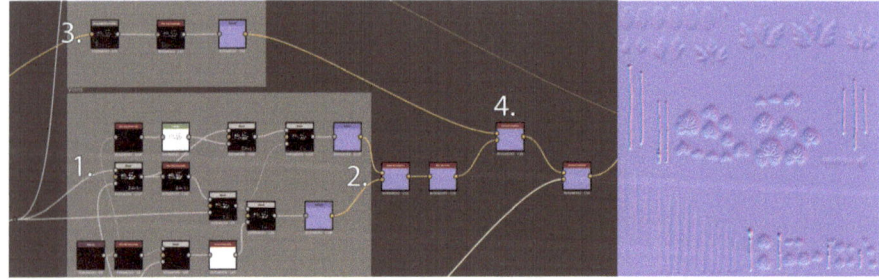

FIGURE 3.12 The node graph for the veins and converted base color part of the normal map, with the normal map visible on the right.

To create a vein mask, you can just use photoshop, create a new layer on top of the Albedo and trace the veins, I use a hard brush and target mostly the primary and secondary veins that I want to get affected, paint the veins white and give it a black background, we will further edit this file in Substance so there is no need to be perfect, as long as everything roughly matches up it will be fine in the end.

FIGURE 3.13   The vein mask used to generate further normal map details.

Looking at Figure 3.12 again:

1. This is most of the network for the Veins. It all starts with the Bitmap node that contains our Vein Mask. This is blurred by 0.6, which goes into a Blend Node where we use the Stem mask created for the Leaf Curvature to mask out just the leaves of the Buttercup and Creeping Charlie, which is then inverted and plugged into the mask input of the next Blend node.

   The Blend Node in slot A holds a slightly blurred (value of 7.18) leaf mask to ensure the veins decrease in intensity over the length of the leaves. This is then plugged into a Normal node with an intensity of 0.8 and plugged into the B slot of the Normal Combine at the end of the graph.

   This gets us a sharper line down the veins, but we need to blend this in a little; the surrounding membrane on the leaves usually folds down a bit when it is close to a vein. We can use our vein mask again but blur it slightly more.

You can see this at the top of the graph. The Vein mask is blurred by 1.03 and plugged into a Levels node. In the levels node, flip the bottom arrows to invert the colors without using an Invert Grayscale; the arrows at the top are squeezed to the right to increase the contrast.

This goes into the blend mask input, masking itself away from the original Blend node created earlier. It is then blended with the blurred leaf mask once again, before a Normal is generated with an intensity of 0.4.

2. Now we have the softer normal generated as well, we can combine these two using the Normal Combine and then a 0.1 blur to eliminate any stepping.
3. The last step in this section is to generate a normal based on the base color, this is the easiest step but will help bring everything together. Start off by converting the Albedo to a grayscale using the Grayscale conversion node, I slightly blur this with 0.1 but feel free to play around with this value and then generate the normal with an intensity of 0.4.
4. All of these are combined in a Normal Combine as well as being combined with the Normals from the previous steps.

With that concluded we can move on to the last step of generating the normal (Figure 3.14).

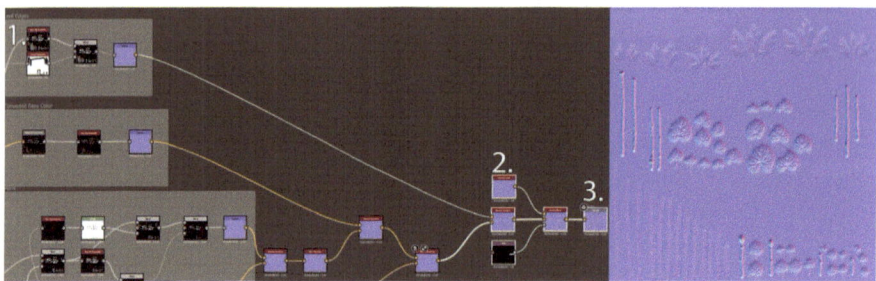

FIGURE 3.14   The node graph for the leaf edges and final composition part of the normal map, with the finished normal map visible on the right.

1. The last step is to create some controls to define the presumed thickness of the leaves; we take the Opacity Mask plugged into a Blur HQ Grayscale. I have set mine to 0.1; keep in mind that larger values here mean the leaf will appear thicker, and smaller values appear smaller; depending on your preferences, I recommend playing around with this value a bit to see what it does. We then use our SVG stem mask to remove the stems from the equation using a Blend node with the mask plugged into the Mask input and generate a normal with an Intensity of 0.3 using a Normal node.
2. We combine our Normal edges with our previously generated normal, but you might have noticed that we have included our flower petals, creating an unwanted.

Normal edge around those, usually flower petals are very thin and delicate so in order to get rid of this, create another SVG node, use the brush tool to paint over the flower petals, and use a Normal Blend node with a Normal Color plugged into Input A, our combined normal into Input B and a flower petal mask into the Mask input.

The goal here is to remove the Normal edge from the flower petals.

3. The combined and final normal is plugged into an Output node with its identifier set to N, Label to Normal and components usage set to normal.

You can now Right-Click in the graph again and select View Outputs in 3D View.

These are all the required steps to generate Normal from texture data rather than high poly or photometric sources. We have set up a graph with individual controls for leaf thickness, curvature, and vein intensity, when moving along in the book, please feel free to adjust values where you see fit; the purpose of setting it up like this is to remain flexible and be able to adjust accordingly, it's best to refer to references when setting these values, but artistic insight is essential as well, if you feel like something looks better with a stronger or weaker Normal, adjust as you see fit.

With all this done, you can Right-Click on the graph in the top left and select Export Outputs; ensure all your final outputs are hooked up to an Output node before doing this. For my location, I used:..*CreatingHighQualityVegetation**workfiles**designer**export*\ *Initial_GroundCover_Atlas.*

For Format, I am using .tga. Targa is a lossless format, which means we lose no quality when saving it to this format. Alternatively, you can use .PNG, but once you start working with Alpha Channels, I find .tga more comfortable. It is also the industry standard file format. Under Pattern, you can alter the default code to ensure your naming matches the identifiers set in the output nodes. I named mine Initial_GroundCover_ Atlas_$(identifier); the "_$(identifier)" part inherits the Identifier set in the output node.

With our textures exported we can now implement these into Blender and move on to the meshing stages.

OPENING BLENDER

With the initial atlas prepared, it is time to start creating meshes. I will demonstrate this in Blender, but you can do it in any 3D software. As a reminder, the intention is to use the meshes to bake down a more optimized texture later, so if you end up not using all the data in the initial atlas, that is perfectly fine. We will make sure to get rid of any unused pieces from the final texture.

If it is your first time in Blender or you are an inexperienced user, I advise you to click around in the software a bit and familiarize yourself with the tool before proceeding.

The chapter will explain some things from the ground up, but it will also assume knowledge of modeling, geometry, materials, and UVs. Global subjects not related to vegetation or the pipeline, some of the things we will set up might not be described in a lot of detail; if at any point you feel stuck with terminology, do not shy away from finding some online resources to explain this topic better.

The areas of interest are marked in Figure 3.15.

FIGURE 3.15   Breakdown of blender UI.

1. The Viewport: This is where we will do most of our editing.
2. The outliner window shows you what objects you have and their names; under the filter, I have turned on Selectable and Disable in Viewport. This allows me to easily toggle between manipulating and leaving alone certain objects.
3. The details panel is where we change all our settings for material or object-specific adjustments, such as the UV channels, materials applied, vertex groups, and material settings.

There is a reference file available that contains all the example files organized in steps in the Outliner; you can find this file here: ..*CreatingHighQualityVegetation\workfiles\blender\Groundcover_Meshing.blend*.

CREATING GEOMETRY FOR THE GRASS AND WOOD ANEMONE

Before we do anything else, I find it essential to import a scale reference; I have extracted the mannequin from Unreal Engine, which can be found here: ..*CreatingHighQualityVegetation\workfiles\source\meshes\SM_UE_ScaleRef.FBX*.

You can import it by going to File, Import, and then Import FBX.

Then, with the scale reference imported, I advise placing it slightly to the side so it does not get in the way when working on the plants later.

You can also Right-Click in the outliner and say Create New Collection. I called mine Utility and dragged the mannequin inside, then flicked off Selectable to ensure it did not accidentally move or get in the way when trying to click on something else.

Then, use Shift+A while your mouse is in the viewport and select Plane. Shortcuts in Blender are specific to which window your mouse is in, so it is important that the mouse is hovering in the Viewport.

Select the plane and hit G to go into Move mode. Select R for Rotate and then *X* to isolate rotation on the *X*-axis. Then, type 90, which should rotate the plane 90 degrees. Alternatively, select Rotate on the left side of the viewport and hold CTRL while rotating to make sure it snaps to 90 degrees. Then, hold the ` key on to your keyboard and select Front to go to the Front Viewport.

With the plane selected, go to the Details Panel and click on the triangle shape; this opens the Data tab. Open up UV maps and click the+Icon to create a second UV channel, double-click the first channel, name it UV0, and call the second UV1. We took all our scans and pictures with us in our initial atlas. Not all these are equally suitable, and while we are building the plant, we will cull some leaves and stems away. We will then re-pack the second UV channel and bake our texture data from UV0 to UV1. This will ensure we do not have unwanted elements in the texture sheet and will allow us to pack it better, increasing texel density and performance.

We can now go ahead and create a material and apply our texture to it, refer to Figure 3.16 on how to set this up.

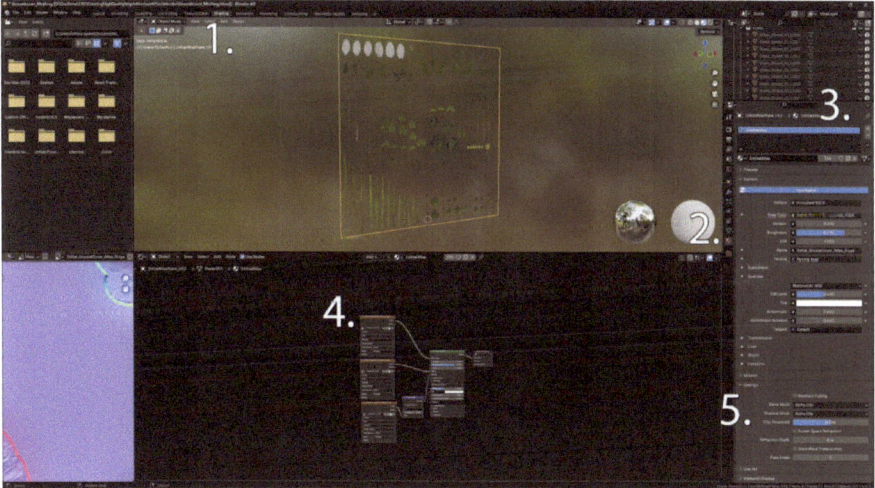

FIGURE 3.16 Screenshot showing how to set up a material and applied textures.

1. At the top of the viewport, go to the Shading Tab.
2. In the details panel, select the Material Tab.
3. With the plane selected, Click the+sign to create a new material I called mine InitialAtlas.

4. Locate the textures exported from Substance Designer and drag them into the material viewport. Connect the Albedo map to Base Color, the Opacity map to Alpha, and for the normal, hit Spacebar and search for Normal Map, add the vector to the graph, and put the Color of the Normal map into the Color input of the vector node, then plug the output Normal into the Principled BSDF Normal input.

5. This will result in the plane turning black, except for where we have textures. To fix this, set Blend Mode and Shadow Mode to Alpha Clip in the details panel.

Now move back to the Modeling Tab, hold ` and get back to the Front Viewport Tab with the plane selected to get into Edit mode and press N to open the Tool tab, under Transform turn on "Correct Face Attributes" and under UVs, turn on Live Unwrap and Keep Connected, This will ensure that if we move vertices around, the UVs will automatically adjust; give it a try and see if it works before moving ahead.

Let's start by cutting out the Grass Strands, refer to Figure 3.17 for the steps taken to cut these out.

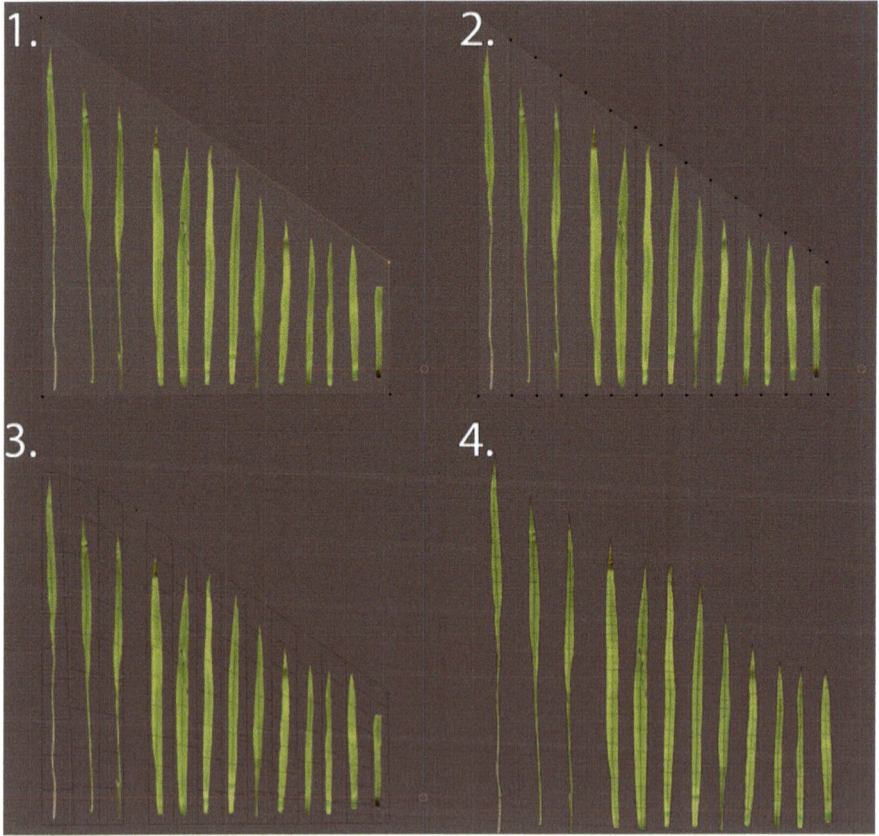

FIGURE 3.17   Screenshot showing the different stages of cutting out grass blades.

1. Initially, I like to isolate the grass stands by moving the four vertices of the plane around the grass blades, this allows me to focus on just these and not worry too much about the other assets, I do recommend duplicating the plane first so you always have a default plane with your texture on there available, you can duplicate meshes using Ctrl+D in Blender.

2. Press Ctrl+R and hover your mouse over the mesh. This will give you a preview of Rings/Loops you can add, if you want to increase the amount scroll up on your mouse wheel or use Page Up and Down, left-click once to move toward the location, then Right-Click to make the edits final.

    You can then go to polygon selection by hitting three select each strand press P and separate By Selection.

3. With the grass strands separated, we can add more geometry to the vertical axis as well. These blades will be bent and twisted later, so it is important to add enough geometry to comfortably do so. However, in this step, it is okay to stay a bit on the low side to just capture the broad shapes. We can add more geometry and refine the shape in the next step.

4. We can now refine our geometry by selecting the vertices and using G to move them around; I have also added an edge in the center to support folding the grass blades later on. Lastly, press Ctrl +. to enter Pivot Editing mode and move the pivot to the bottom of the grass blades.

When building meshes for vegetation, there are a couple of things to consider. First and foremost, how will we manipulate meshes later? If they are bent, they will need enough geometry; second, what is our mesh and opacity ratio? Triangle count impacts performance, but in many instances, overdraw impacts it even more, so it is, in most cases, better to add more geometry and closely follow the shape of the texture to eliminate as much overdraw as possible.

But now that Unreal Engine has Nanite, we are moving toward a world where Opacity maps become more expensive than millions of triangles. At the time of writing of this book, this new standard of triangle counts is not yet defined and solidified, so the book focuses on a hybrid approach where there are enough triangles to eliminate most overdraw, but in some cases, still relies on the opacity map for further detailing, with this knowledge you will be well equipped for both worlds.

For good measure, select all the grass strands, then Ctrl+A, and select Rotation & Scale Transforms. In the bottom left corner, make sure Location is unticked to keep the pivots in the same place you put them. This will ensure the meshes are clean of any rotation and scale values that can get in the way later.

Since we will use these elements to create our completed strands and clusters later, it makes sense to think about our Level of Detail meshes in this step (or LODs for short). LODs are a cheaper version of the mesh that will be swapped out in the game engine at a set distance, so if, for example, our first LOD (LOD0) is twenty thousand triangles, this will get swapped out at, for example, a distance of twenty meters to a mesh that is ten thousand triangles. This is a way to save resources, as objects in the distance do not require the same level of detail as objects close to the camera. A common number of LODs is 4, with a roughly fifty percent drop in tri count per LOD.

With that in mind, it is important to adopt a slightly counterintuitive mindset because, as a rule, the most viewed LOD will be LOD1, and LOD0 will only be visible close to the camera. So, to improve the whole image, it is often better to focus on a good-looking LOD1 than to spend all your time and effort on LOD0.

This does not mean LOD0 should not be the best-looking LOD up close, but I recommend avoiding the thought process of "LOD1 is less important because it is further away." Every LOD is important for its respective distance, and they should all be carefully crafted.

In many instances, LODs can be automated by third-party software or the engine itself. You can also rely on decimation tools or modifiers within your 3D software package. However, in the case of vegetation, where we rely on Opacity masks to hold the shape, these tools usually remove the geometry that is needed to keep the overall silhouette intact. Therefore, in almost all cases, it is worth the time and effort to make your LODs by hand.

With that in mind, creating them is a time-consuming process, and if you are reading this book to learn about vegetation, your time is not well spent working on these for too long. If that is your case, I advise you to build LODs for only one asset to go through the process and build just LOD0 for the others. However, if you intend to use your assets in a real-time environment, it will be essential to create them all.

Refer to Figure 3.18 to see an example of how I set up my LODs. In this step, these are just estimations, and they might be shifted around a little when bending, twisting, or seeing the final asset.

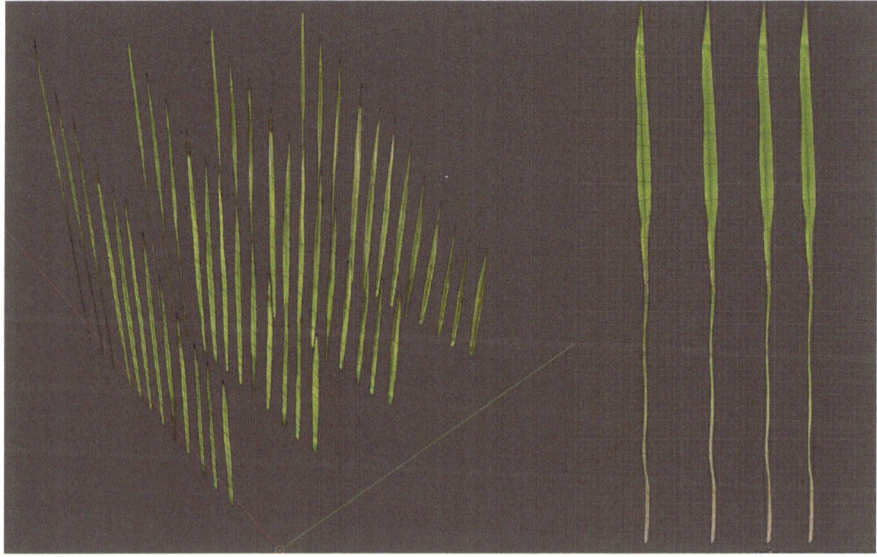

FIGURE 3.18   Screenshot showing an example of LODs for the grass strands.

Since we will set up an extensive pipeline, it is worth naming your pieces. I have called mine Grass_Strand_xx_LODx, where the x refers to a number. Additionally, I placed them in a Collection called Grass – refer to the reference blender file if you are unsure what to do.

CUTTING OUT THE WOOD ANEMONE

The process of cutting out the Wood Anemone is very similar to that of cutting out the grass strands, but there are some things to be aware of. The following explanation will focus on the unique aspects. For everything else, use a similar workflow to that of the grass.

For the flower bud, we will use a 3D mesh. For longer distances, it would be better to use a card or a crosscard (two cards rotated 90 degrees from each other so they are visible from two angles), but for the sake of the book, we will focus on quality when making close-up shots a 3D mesh holds up much better than a crosscard and considering the way the industry is going at the time of writing this book, opting for more triangles to increase quality is the way to go.

Refer to Figure 3.19 for explanations of how to approach the modeling of these two elements.

FIGURE 3.19 Screenshot of the blender viewport showcasing different steps taken to model the flower bud and leaf of the wood anemone.

1. From left to right, start out by roughly cutting out the shape, adding geometry using Ctrl+R to add loops, and closely follow the shape.

    There is quite a large chunk sticking out on the right side. If we keep it in, this will become a repeating element, so I opted for cutting it out, making the shape more symmetrical and less recognizable. That way, if we copy and paste this around over a large number of flowers, the shape won't stick out too much.

to closely follow the shape of the flower bud, we need a lot of geometry, but this is mainly required in the silhouette so we can collapse some vertices in the center. To do this, select the vertices you want to merge, Right-Click, Merge Vertices, and select Collapse.

We can then add a Mirror Modifier in the Modifier Tab of the Details panel. This is harder to see in the front view, so refer to the outlined image on the top right in Figure 3.19. Set the mirror to the corresponding axis (in my case, Y) and pull out vertices until you get a rounder shape. Apply the Mirror modifier once you are happy with the result; I have then relaxed the shape a little, making it more round in general, and applied a Decimate modifier to remove any unnecessary vertices; I have set the Decimate modifier to 0.5.

2. When we deform the leaves later, we want to do so the same way as seen in the references. Therefore, we will need geometry in the right places, especially the veins highlighted on the left. Until now, we have only dealt with straight leaves, but the WoodAnemone somewhat splits into three main veins.

3. I find the easiest way to achieve this is by modeling the center part first, making sure the center edge follows along with the vein of the leaf, and making sure we have some geometry in the area where the leaf splits off.

4. Then select those edges and use shortcut E to extrude them. This will create additional geometry that we can then edit to wrap around the split-off leaves, again, making sure the center edge follows the vein, as seen on the texture.

Follow similar steps for all other leaves until you end up with a kit like Figure 3.20.

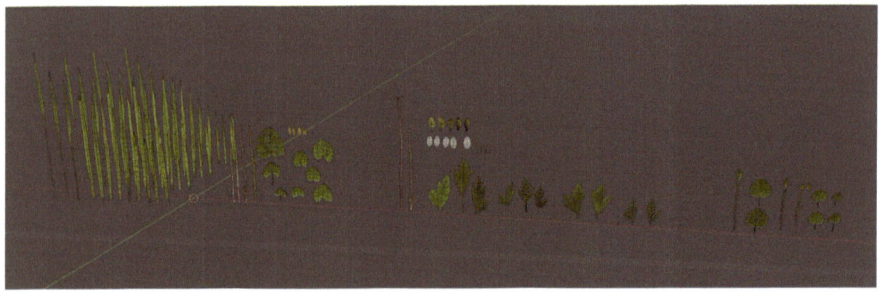

FIGURE 3.20 A screenshot showcasing the final set of geometry we will bring to the next step.

A couple of things to note are that I have only prepared LODs for the grass. We will cover setting up the LODs in the book, and the process is identical for all other plants. You might notice that not every leaf and stem has been modeled out, which is intentional. In the next step, we will rebake the atlas to improve the packing, but it is also an opportunity to eliminate unwanted elements.

The leaves that I have culled are leaves that still had too much of a specular response in their color after our fix in Substance when building the maps, and I have also removed leaves that were too distinct or similar; when repeating plants, it is best to use textures that do not stand out too much, this is to avoid repetition, but additionally, when packing later we want to give pixels to the ones that matter the most if we have two very similar leaves it is better to give one of them twice the amount of texture space opposed to bringing both of them to the next step, that way we maintain the highest texel density possible.

Texel density refers to the number of texture pixels (texels) per unit of 3D surface area. And generally you want to keep this somewhat uniform across all your assets; a good goal would be to have at least 1024 pixels per meter.

As you get more experienced building plants, you will find it easier to make this selection early on, which is something we could have done – but the workflow shown in the book is a workflow that allows you to see things in your 3d software, play around with leaves to see if they work the way you want them to work, and then remove them at a slightly later stage if time allows this is my preferred method as it reduces the risk of culling too much too early on and finding myself with a very minimal kit to work with when building the plants later on.

I have also gone ahead and made the meshes for the Creeping Charlie and Buttercup.

REBAKING OUR INITIAL TEXTURE INTO THE FINAL TEXTURE ATLAS

When rebaking the atlas, there are some initial considerations we need to make; for simplicity's sake, I will rebake all the plants into a single atlas. Again, this is efficient if you intend to use all plants simultaneously as they will share one material that way the engine will be able to batch them up and render them in a single draw call.

A draw call is a request from the CPU to the GPU to render a game object on screen; a lower amount of draw calls means better performance; therefore, we always try to build things in a way that allows us to keep the number of draw calls low. If you intend to use them in isolation or are unsure about the use case, it is most likely best to split it into four atlases. When getting into optimization, you will find there is usually no right or wrong, and the solutions and considerations will not work in every single situation, so use your best judgment and adjust accordingly.

We can start by opening the UV Editing tab, selecting all our meshes, and going into edit mode. Then, press A to select all elements and have them show up in the UV editor. At the top right of the UV editing tab, there is a dropdown menu that should show you two UV channels, UV0 and UV1. If you have not created a second UV channel in previous steps, you can still do so now; it will copy whatever you have in your first UV channel.

Refer to Figure 3.21 to know what you should be seeing. The areas of the texture that do not have UVs are the leaves that I ended up culling away, and the goal is to remove these from the texture now.

Make sure you select the second UV channel in the top right of the UV editor, in my case it is called UV1, and with all the UVs selected, go to UV, merge, and choose by

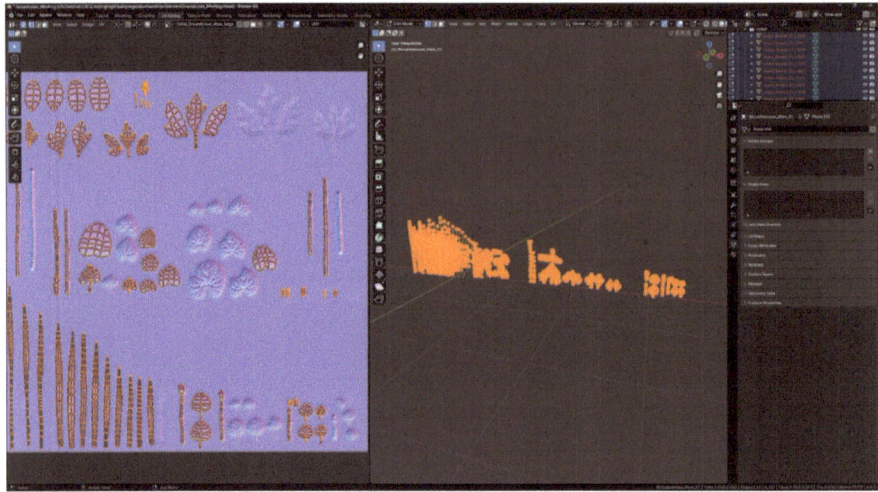

FIGURE 3.21   The UV editing mode in blender with the UVs visible on the left and the viewport on the right.

distance. Set the distance to 0.0001; this is to ensure no vertices have become disconnected during the modeling process and will be seen as separate UV islands.

We have some tiny UV elements in our atlas, so now it is a good idea to check those to see if you have any unwanted merges; if that is the case, lower the Merge Distance even further.

Assuming everything is merged correctly, you can select all the UVs again, go to UV, and then Pack Islands. Set Shape Method to Exact Shape (Concave), and make sure you turn rotation off. We have already created a normal map. If we rotate that now, we will mess with the direction of the normal map, resulting in inaccurate lighting calculations.

The Margin Method can be set to Scaled with a margin of 0.004. This will ensure comfortable space around the UV islands to avoid bleeding but still result in tight packing. Turn on Merge Overlapping. If we do not turn this on, it will pack the Grass LODs as separate islands. Since we intend to reuse the same texture across our LODs, we want the overlapping UVs to stick together. Lastly, set the Pack to the Closest UDIM and press Ok.

You should now see a newly packed UV and no visual change to the meshes. This is because we are applying the texture to UV0 and have edited UV1. Refer to Figure 3.22 to see the intended packing and settings. Depending on your meshing, this could look different, and that is okay. As long as all the elements are re-distributed and fill up most of the square, you are good to go ahead.

Please select all the meshes and go to File, Export, FBX, and export all meshes into a place that makes sense to you, I have exported mine here: ..\CreatingHighQualityVegetation\workfiles\blender\export\UVTransfer\Groundcover_Meshing_UV1Packed.fbx.

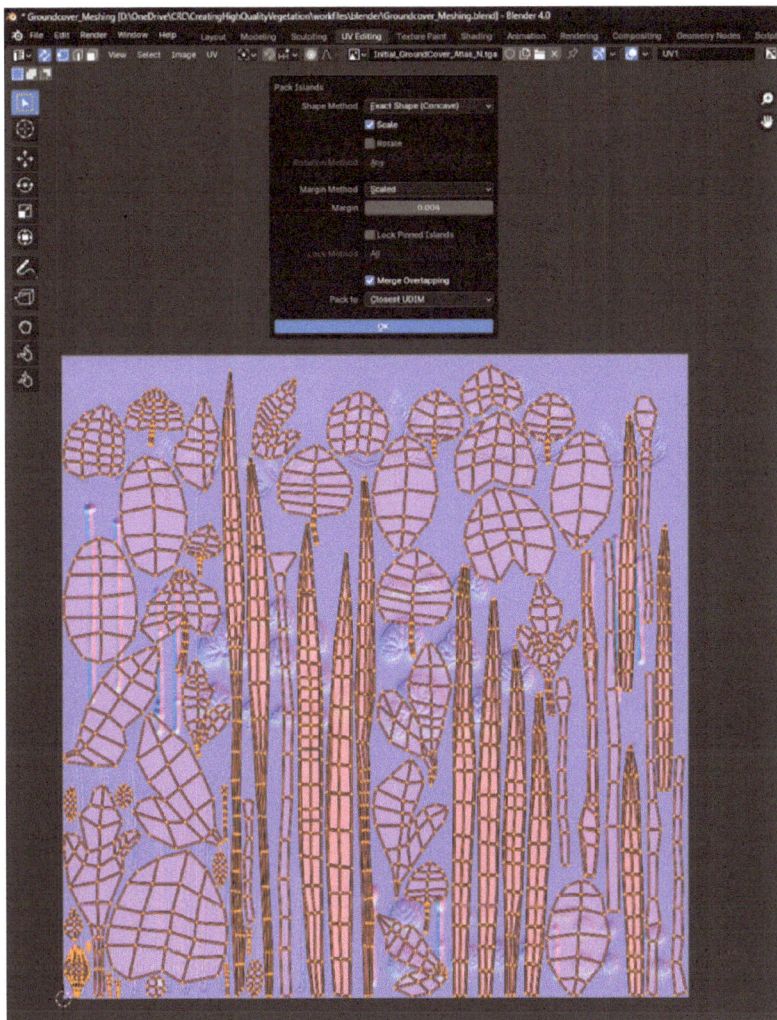

FIGURE 3.22   The packing settings used alongside the newly packed UV1.

Then go back to the SBS_GroundCover_AtlasProcessing substance file and drag and drop the mesh into the Resources folder. If it asks you if you want to import it as a UDIM Mesh, select No.

A UDIM mesh is a mesh that has multiple UV layouts and requires a texture for each, this is not commonly used when building vegetation and can be ignored. Inside the resource folder, you can Right-Click the mesh and say Bake Model Information. This will open the baking window.

Firstly, set the folder to something that makes sense to you. In my case, I saved it here: *..\CreatingHighQualityVegetation\workfiles\designer\export\T_GroundCover\Bakes.*

And set the name to something that indicates these are the rebakes: T_GroundCover_ Rebake_$(bakername). In this case, the baker name will be set to whatever we name the baker in a later step.

We can now move on to the Bakers Default Values. For the size, you can set this to 4096; if you have a slow GPU, you might want to set this to a lower value for the first couple of bakes, then shift it to a higher resolution for the final bake. This will speed up iteration, especially if you're baking actual mesh data; for now, we will only transfer texture data, so the bake should be very quick regardless of resolution. The format will once again be set to .TGA.

Click the Add Baker button and select Transferred Texture from Mesh. Do this four times as we will be transferring four textures: the Albedo, Normal, Roughness, and Opacity map. This is where the baker's name becomes important. I have named mine C for Color, N for Normal, R for Roughness, and O for the Opacity map.

In each of these, select its corresponding texture file previously saved here: ..\CreatingHighQualityVegetation\workfiles\designer\export\Initial_GroundCover_Atlas.

If you have used your own folder structure, these are the texture files of the initial atlas. For the Normal baker, turn the Normal map checkbox on.

In the baker list, there is a column for UV set; this one must be set to 1* for all the bakers, and the Format is set to .tga.

The dilation width should be set to 128, and we should turn off "Apply Diffusion" and "Average Normals."

These are a lot of settings, so make sure to double-check yours. Alternatively, you can use the preset found here: ..\CreatingHighQualityVegetation\workfiles\designer\ UVTransferPreset.json.

You can load a preset in the bottom left of the baking window, and if you use that preset, it will be set up identically to mine.

Refer to Figure 3.23 for the final setup of the baking window.

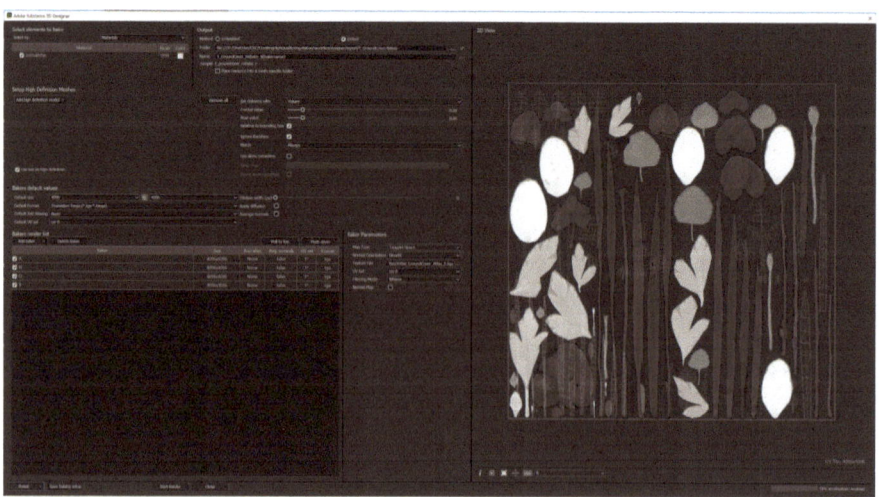

FIGURE 3.23 The baking window with all the settings put in.

Once you are ready, you can press Start Render, which should give you a new texture that matches UV1. The textures are then placed in the Substance file Resource folder.

Then, Right-Click on the .sbs file in the Substance Explorer and click New Substance Graph to create a new graph. I have called mine T_GroundCover. Copy the output nodes created previously or create four new ones corresponding to their inputs.

For performance reasons, we will also do some Texture Packing. Texture packing means we will pack grayscale textures into a channel of a different texture. In our case, we will pack the Opacity together with the Albedo. To do this, put down an Alpha Merge node and plug in the Albedo and Opacity map.

Refer to Figure 3.24 for the correct setup.

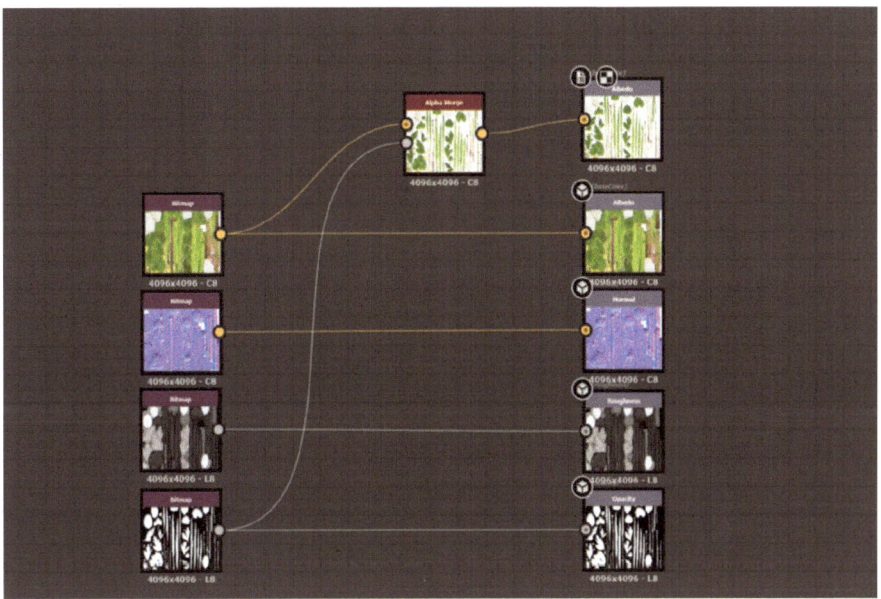

FIGURE 3.24 Simple substance setup to re-export the textures.

Right-Click on the graph and select Export Outputs; these are the final textures we will use in Unreal Engine so ensure they use the Unreal naming convention; if you have followed the steps in the book, this will happen automatically; if not, you can name them manually on export to something along the lines of T_AssetName_C the prefix is for what type it is, in our cases a Texture, the suffix stands for which texture is inside

C = Color
N = Normal
R = Roughness
O = Opacity

For our packed texture we can use T_AssetName_CO to indicate it contains two types of texture.

I have saved mine as .TGA in this location: *..\CreatingHighQualityVegetation\ workfiles\designer\export\T_GroundCover*.

We can then open up Blender again and replace our previous textures with the ones just exported; this will initially appear broken as the texture is being applied to UV1, so we need to make sure our meshes have the correct UV Channels only. To do this, go into the Data Tab where we previously created a second UV channel, and remove UV0, you could opt for renaming UV1 to UV0 as well, but we will not be touching the UVs again, so I will leave mine as is. Unfortunately, there is no easy way to remove all the UV0 channels of all meshes in one go, and you have to do it for each mesh.

If you have made it this far, congratulations. The more complex technical setup is now done. We have successfully packed our final texture and mapped it to the correct UV channel. We can now focus on the fun part of building the assets.

If you have gotten lost or stuck along the way, you can use the 2.Reproject UV0 to UV1 Collection in the example Blender to continue from here.

CREATING GRASS CLUSTERS

The first thing I do when doing anything is to have a good look at all my references.

I cannot stress enough how important this is; a teacher of mine once said, "*Your work can only be as good as your best reference*," and it has stuck with me ever since.

I have seen so many assets that are close to good but fall apart due to the artist not having used or looked at references. In this case, I have been able to shoot the reference myself, so I picked up exactly what I needed, including individual clusters, but sometimes, when building vegetation, it can be hard to see what is going on. What helps me in these cases is to paint and do an exercise where I draw over my reference; this usually allows me to judge the shape and what I will build better. It's not rocket science. Just pick up a pen and trace over the shapes you are seeing, and as you go, think about what you are drawing; how do the grass strands intersect? Are they bent or twisted; how do they spawn off the stem? Remember the phyllotaxy? In the case of our grass, we will be going for an alternate phyllotaxy. When looking at references and combining it with a paint over you are working with different parts of your brain, and in my opinion, the reference sticks better, and I am filling my brain with information about the plant I am trying to build.

Additionally, this will create an abstract sketch of the reference, allowing you to look at it with less noise and seeing some of the shapes for what they are, you can do the same exercise on your final asset. Then, instead of comparing the asset to a reference, you can compare your abstract to your abstract on the reference. This makes it much easier to see if you have settled on the correct length and size for the leaves, if they spawn at a good distance from each other, and if you have gotten the folds and bent right.

I recommend anyone reading to try this method and see if it works for you.

When looking at the grass and many other vegetation assets, for that matter, you will occasionally see a life cycle; I scanned my assets in early spring, so it is not that visible, but you will find some dead strands, broken strands, or strands eaten by bugs, for example. For the sake of simplicity, I will be covering only the healthy aspects of the plant, but if you are looking for a way to expand your knowledge after reading the book, adding a very clear lifecycle to your assets is a great place to start.

For the grass, I have some great reference photos that you can find here: *..\CreatingHighQualityVegetation\ref\ref_grass*

Alternatively, these are the photos in the source files as well, I just cropped out any unnecessary elements. These references will work as great starting points for our modeling, but we have to keep in mind that they are no longer in their natural habitat, due to the absence of water, they have gotten a bit softer, and they are lying down, not standing up – so take them as a reference but look at the close-up shots for a more accurate representation of the grass in its natural habitat.

You can import the provided reference by going into a front view and dragging an image into the Blender viewport, and this will automatically create a plane with the image applied as a texture.

Now is a good time to bring in a scale reference. If you have not already import the one provided here: *..\CreatingHighQualityVegetation\workfiles\source\meshes.*

Scale the references and grass strands to the correct size. This will help us a lot in the future. Refer to Figure 3.25.

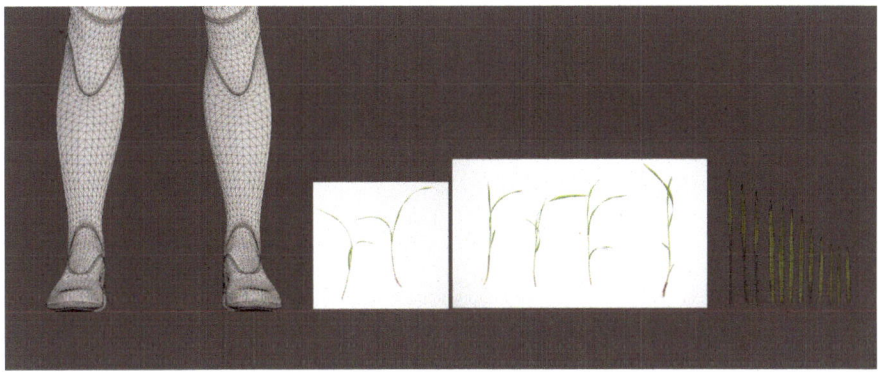

FIGURE 3.25   The blender viewport showcases the scale reference, the grass reference, and the grass strands.

Once you have scaled them, the meshes will contain transformation data. This can get in the way when we want to add modifiers to the meshes later, so select all meshes and use the shortcut Ctrl+A and click Rotation & Scale. There is no need to reset the location, as this will move the pivot to World Origin (0,0,0 or otherwise known as the middle).

We can now move on to adding some modifiers so we can non-destructively edit the meshes according to our needs. We will add a fold modifier and a twist modifier; however, we have some strands that include a stem, as seen in the references; the stem does not fold or twist, so we want to exclude these from the modifiers we add. In Blender, the way to do this is to apply Vertex Groups, select a grass strand with a stem, go to edit mode, select the vertices of the grass blade part, switch to the Data Tab in the details panel, the same tab where we added a second UV channel earlier. But this time, click the+Icon under Vertex Groups, call the group "Blade," and then click "Assign." This will allow us to isolate this part later in a modifier.

Select the grass strand and go to the Modifier Tab, select Add Modifier and click Simple Deform; you can then choose to rename this for easier recognition; I have called mine "Fold," set it to Bend, and depending on your Pivot orientation, this might differ, but I am bending it over the Z axis. Open the Restrictions Tab and select the Blade Vertex Group. If you play around with the angle now you should see only the blade move. Refer to Figure 3.26 to see what is expected; the modifiers in that figure are exaggerated for demonstration purposes.

FIGURE 3.26   Blender viewport showcasing the modifier tab with two simple deforms applied, one for the fold and one for the twist.

We can then select the other two strands that have a stem, double-check that they also have the Blade Vertex Group, and select the strand with the modifiers last so it is the active object (you can see if it is the active object if the orange outline is brighter when you select it) then use Ctrl+L and click Copy Modifiers, this will apply the same modifiers to the other strands as well.

For the remainder of the grass strands, we will not need the restrictions to be applied as they do not have a stem but you can go ahead and copy over the modifiers, then remove the restrictions.

I have opted not to twist them too much, as we will be able to influence the twisting of the mesh when we deform them with curves in the next step.

Like how we create a mesh, you can use Shift+A then select Curve, and select Bezier to create a Bezier Curve. In Edit mode, you can then move the points around. A Bezier curve has handles, and since the object we have is quite small, we will need to adjust these handles accordingly. With a point selected, you can use S to scale them down. Alternatively, you can select both points, click V, and set them too automatic.

Place one side at the top of the strand and one at the bottom, then with both points selected click F to subdivide the spline and create an additional point in the middle. I recommend naming the curve like the mesh and adding _Curve to make it easy to recognize later; I usually place my curves and meshes in different collections as well; this way, I can easily make one or the other unselectable which will help us later. Select the mesh and add a curve modifier, set the deform axis to Z, and select your Curve in the Curve Object dropdown menu; refer to Figure 3.27 to see the final setup.

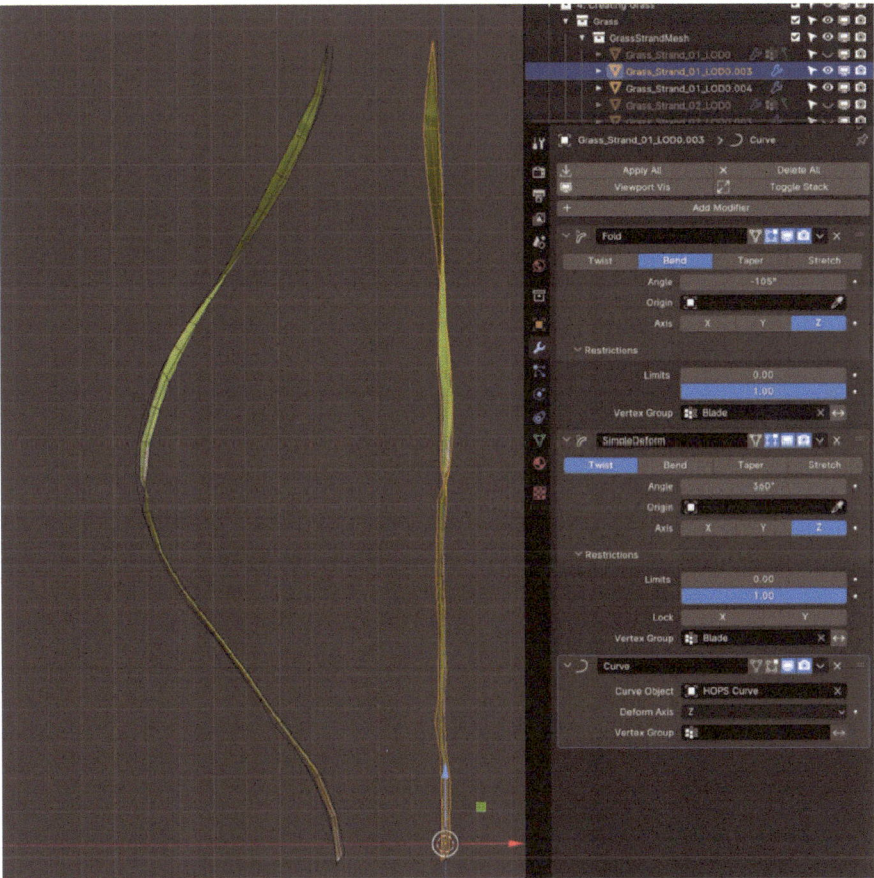

FIGURE 3.27 A mesh with a curve modifier the left has had its center point moved, the right is the default state.

What this allows us to do is move the spline points and deform the mesh accordingly, if you have never worked with splines before, a Bezier Curve can be a bit intimidating; if that is the case, I encourage you to take fifteen to thirty minutes now to just play around with the curves, until you get the hang of how they work, here is a list of all the features we will be using to model the rest of the grass.

Ctrl+T=Adjusting the Tilt or Twist

Alt+S=This will scale the spline knot and the mesh accordingly, you can use this to make the mesh thicker or thinner in specific areas

R=This will rotate the handles and the spline knots, and the mesh will deform accordingly

F=If extra control is required, we can select two knots and use F to subdivide. We can then use the extra point to further deform the grass blade.

In some cases, depending on the position of the spline knots you will find your mesh jumping around a bit, twisting into odd positions. To fix this you can use Ctrl+T to move it back into the correct position. We can turn on the Curve Normals to more easily see what direction the curve is going. To view the curve normals, go to the top right of the Viewport and select Mode Overlays, then turn on the Normals; because our grass blades are so small, you might have to edit the length to something like 0.01 you can refer to Figure 3.28 to see where this is located.

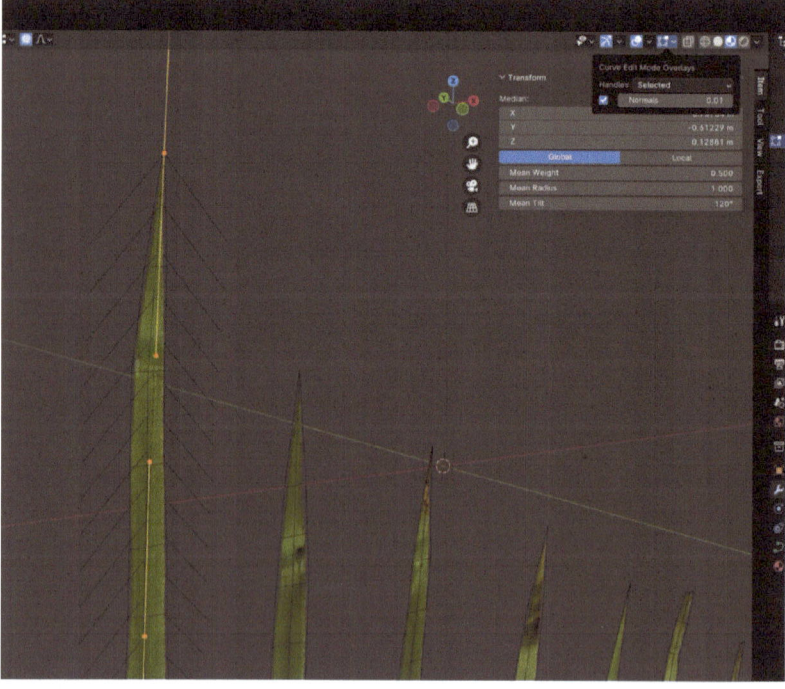

FIGURE 3.28 The top left of the blender viewport showcases where to find the curve normals overlay.

Now, we can start modeling the strands to use them to create clusters later. Since the Curve Modifier has been added, we need to move the mesh and the curve together; otherwise, the mesh will deform unpredictably. Additionally, if you want to reuse a strand, you need to duplicate both the mesh and the curve, to keep things simple, I would use Shift+D on both and then Link the object data by selecting both meshes, clicking Ctrl+L and select Object Data, that way when you make edits to the mesh it will be applied to all meshes, but the curves can be edited independently from each other.

If you have the references in your scene, you can trace them as well as possible. This does not need to be perfect, but I find tracing them over the reference to be an excellent starting point; the important part here is that we get something on our canvas we can judge, maybe they are too big or too small, too bendyin or too straight but we will not know this until we have something to look at, so do not overthink this step, do something to the best of your ability.

We have to keep in mind that these references are taken after they have been removed from the ground, so their shape has slightly changed. Additionally, since they are lying down, you will notice in Figure 3.29 that I have not traced them exactly as the reference showed. Instead, I chose to make all the strands appear more upright. This is also something you can observe in the main PureRef file.

FIGURE 3.29   The grass strands seen from a front view.

If you have completed the paint-over exercise, pay close attention to the direction of the grass strands. I sometimes draw arrows to determine the ratio of strands that go upwards, sideways, and downwards. This helps me understand and accurately depict the orientation and flow of the grass.

After you are done tracing the meshes in the 2D view, please switch to the 3D view and ensure you move them along all axes. This step is crucial in ensuring there is depth to the strands, which will significantly enhance realism and ease of use when we create the clusters later. By adjusting the strands in the 3D space, you add the necessary volume and dimensionality, making your model look more lifelike and dynamic. This added depth will not only improve the visual quality but also facilitate better clustering and arrangement in the subsequent steps.

An important thing to factor in here is from which angle the plant will be seen. This depends on your specific use case. Since this is ground cover, it will be low to the ground, so if you are doing this for a first- or third-person game, you want the volume to be seen from above. This does not mean that you need to rotate all the strands up so the thicker side faces up, but the majority of the volume should be seen from the primary viewing angle.

If you refer to Figure 3.30, you will see what I mean, notice that the camera is tilted to above, the strands have been moved into all three directions, and the strands are twisted using Alt+T to primarily be visible from above.

FIGURE 3.30   The grass strands seen from a perspective view.

Once you are happy with your result, we must collapse the meshes for the next step. However, we would like to maintain this version in case we need to make any edits, so I strongly recommend you select all your meshes and curves, select Shift+D to duplicate them, and move them slightly behind the original.

Before we can join everything together, we need to apply all our bend and twist modifiers. The easiest way to do this is to use the Modifier Tools in Blender; however, this is not enabled by default, so we will need to activate it ourselves. In the top left, go to Edit, Preferences; in the Blender Preferences window, go to Add-Ons. In the Add-Ons search bar, look for Modifier Tools and enable it.

Now, if you go to the modifier Tab of a mesh, you will see four new buttons: Apply All, Delete All, Viewport Vis, and Toggle Stack; for our purposes, we are after the first. So, select your duplicate meshes and click Apply All. Now the link to the duplicated curves is broken, so we can remove those to clean up the scene a little. Again, if you have put your meshes and curves in separate collections, you can quickly make all your meshes unselectable to give yourself an easier time clicking on the curves.

For good measure, select all the meshes, Ctrl+A, and apply the rotations and scale once more, then select all the meshes in a strand. Join them together using Ctrl+J and adjust the pivot using Ctrl +. so it sits neatly at the base of the strand. Give them a name that makes sense to you, I have named mine "Grass_Strand_Cluster_06_Master" I called them Master, so I know that this is the combined base mesh, I do recommend keeping your scene organized. If we need to make changes to the final mesh later and it requires us to make changes in the initial steps, it will be much easier to feed our changes through to the final meshes.

We will need to do some preparation for the wind shader we will set up in Chapter 11; Vertex Colors will drive this and will need to be set up now.

WHAT ARE VERTEX COLORS?

Essentially, every vertex on a mesh contains data. Vertex colors are part of this data and contain a number from zero to one in each channel (RGBA). This data can then be used within game engines to drive specific effects.

In the case of wind, we can, for example, define that for the red channel, we apply a value of 0 at the stems and have a gradient toward a value of 1 at the tip of the leaf. We can then tell our wind material to look at this gradient and say we want you to have no effect at 0 and maximum effect at 1. Isolated per channel, this will generate very simple results, but as soon as we start layering multiple vertex colors with individual effects, this data set becomes one of the Vegetation Artists' most powerful tools.

To add vertex colors, we need to isolate our Master clusters. You can do this by selecting a mesh and using Shift+H this will hide all unselected meshes, and you can use Alt+H to make everything visible again. Keep in mind that in Blender, you can only add vertex colors to one mesh, so you cannot select them all and batch-edit them. This is a bit inconvenient, but it does remove some room for error.

With our grass strands isolated, we can select our first mesh, and in the top left of the Blender viewport, switch from Object mode to Vertex Paint mode. Refer to Figure 3.31 for a brief explanation of where things are located.

1. This is where we change which color we want to paint. We will be focusing on pure Red, Green, and Blue values
2. Some brush settings, but more importantly, the blend mode, we will be switching between Mix and Add
3. The different tools available to us in Vertex Paint mode
4. This is the brush used to pain. You can grow or shrink the brush using the radius settings highlighted in point 2

As usual, if this is the first time you are working with vertex colors, just make a duplicate of the mesh and play around with it for a while; the individual steps are not complicated, but it can take a couple of tries to get comfortable with them.

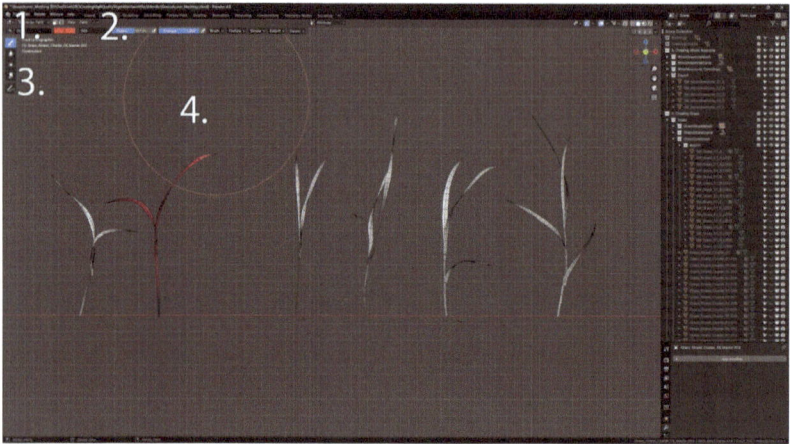

FIGURE 3.31   The grass strands isolated and blender switched to vertex paint mode.

The shader we set up in Chapter 11 will use the vertex channels for the following purposes:

Red

Wind Intensity. We want the wind to affect the part that is supposed to be rooted in the floor, so for that reason we will apply a red color only to the tips.

Green

This will affect the speed at which the wind moves, so we can opt to paint a brighter value at the tip, or we can make every strand a unique green color to break up the movement and make it look more realistic by having them move at different speeds and timings.

Blue

In addition to the detailed wind controlled by Red and Green, we will build a general swaying effect; Blue will offset this effect so we can have each strand move at a slightly different time.

Refer to Figure 3.32 for a breakdown of the steps taken to achieve this.

1. Selecting a black vertex color and pressing Ctrl+X to fill this up
2. Set the paint mode to Mix (this is the default setting) and paint the tips of the leaves red. Then, use the Blur mode to smooth out the transition.

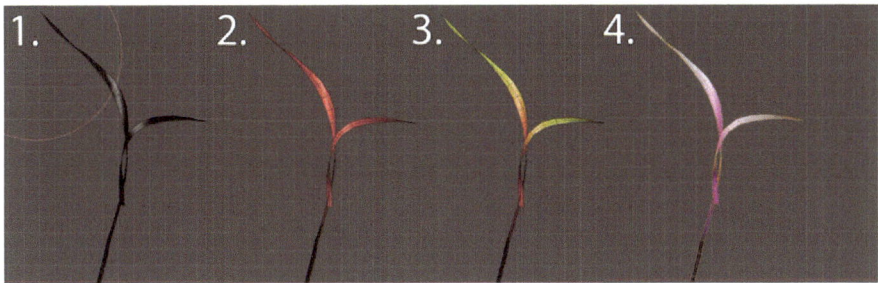

FIGURE 3.32   Break down of adding each individual vertex painting step.

Alternatively, you can fill up the whole mesh with Red using Ctrl+X and paint the bottom black.

3. It is important to now switch the paint mode to Add to ensure we are not replacing our red with Green but instead Add Green to the Red. The green will offset the speed so painting it at the tips will make them move slightly faster, you can also paint the strands two different values, like 0.8 and 1 to make them move somewhat different from each other. Make sure you do not do this where they connect to avoid having them disconnect in the engine later.

4. Lastly, select a blue color between 0 and 1 and with the paint mode set to Add paint the whole mesh with that color. When doing the other strands, pick a different blue color so they will sway with an offset.

Repeat these steps for all strands you have made; if you are having trouble figuring this out, I want to point out that the wind will only be relevant if you intend to make a video or real-time environment; it won't be visible on still images. On top of that, the shader we will set up will be partially driven by Unreal Engine, so even without the Vertex Colors, you will be able to set up a decent-looking Wind Material; in short, if this step proves troublesome, feel free to move on for now and come back to it later.

That said, it is very important for every vegetation artist to get a thorough understanding of how to apply and work with Vertex Colors.

This completes our technical setup; keep in mind that if you are using LODs, you need to add vertex colors to them as well so the wind will not stop moving when switched to a lower LOD.

To create the LODs, Blender has some handy tools to maintain the same setup and just switch out the modifier data. I usually create something I refer to as my LOD factory in Blender. It is just a separate space where I do everything LOD-related. Refer to Figure 3.33 to see what that looks like.

1. Here all the LODs are visible, in the screenshots have already been relinked.
2. The LOD strands we have built before, in the screenshot we are seeing LOD1, LOD2 and LOD3.
3. The outliner which I use to identify which strands to relink.

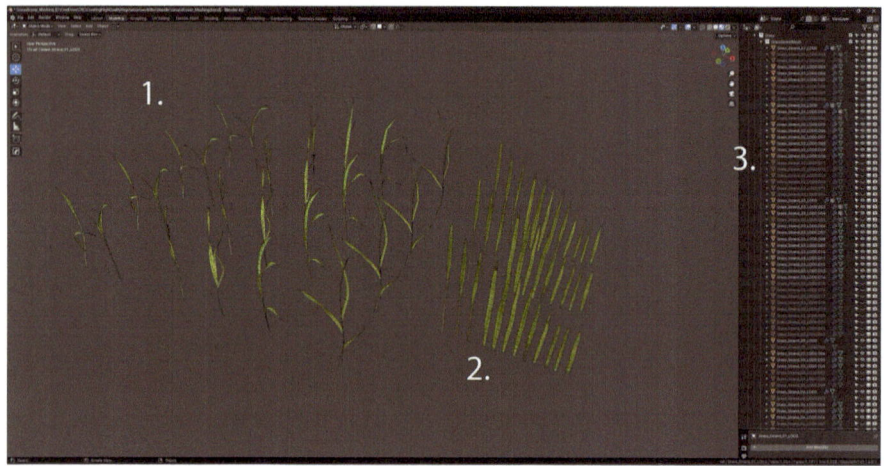

FIGURE 3.33   A setup where I can create all my LODS.

In order to create our LODs, you will need to duplicate the grass strands that still have the curve modifiers; this is why I recommended making a duplicate of those before doing the vertex colors. Move the duplicate slightly behind the original, Referring to Figure 3.33, select the strand you want to replace in Section 1 mentioned in Figure 3.33. Then, select the strand you want it replaced by in Section 2; with both selected, click Ctrl+L and say Link Object Data; this will relink the mesh to a lower LOD – Repeat this step for every strand and every LOD in your LOD Chain.

With our LODs created we can now use these strands to create our clusters.

When creating clusters, it is important to bring up our reference again and double-check that we are creating what is expected and seen on the reference; I like to create clusters of different sizes; the larger the cluster, the better it is for performance, but smaller clusters allow for more flexibility in the engine. This is because larger assets fill up spaces with less objects used. When referring to size here, I am talking about the footprint of the asset, not the scale.

Creating clusters can be as simple or complex as we make it. I like not to overthink this step, as never in my career have I built something the first time and agreed with the visuals. It is all about iteration, both in Blender and in the engine. Usually, when imported into the engine, I find it is too small or too big or there are too many recognizable elements, so doing something quickly and seeing it in engine sooner is better than going all in on the details at this point.

Refer to Figure 3.34 to see a step-by-step approach on how I approach building clusters.

If this book should teach you anything, it would be to look at references. So once again, open up your reference board, break it down, and use the information you extracted to inform decisions when making these clusters.

The main elements from the reference I am capturing with these clusters are that they are slightly higher in the center and scale down toward the edges. Additionally, toward the edges, the grass seems to point more toward the outside than the top.

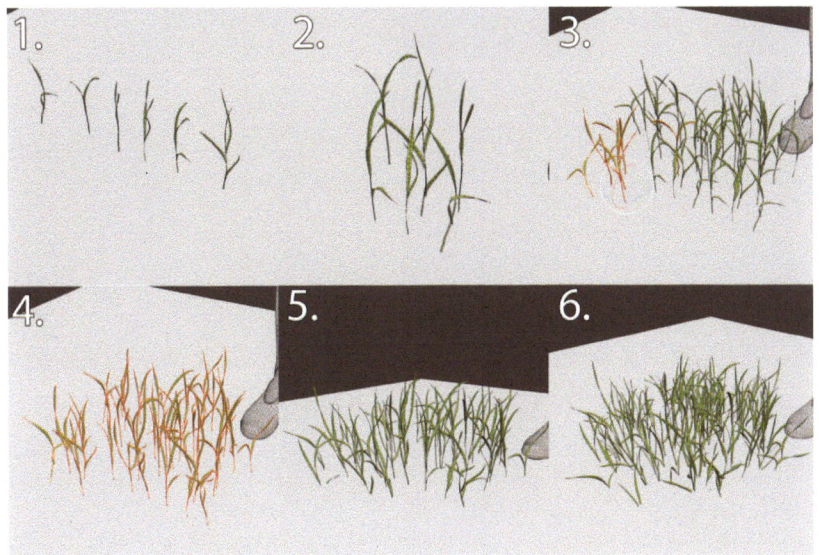

FIGURE 3.34 Image showing multiple steps on how to create grass clusters.

1. Prepare the strands you intend to use for these clusters; the large ones might not be immediately suitable for smaller clusters, so feel free to add or remove to this kit as you see fit.
2. Group them up without too much thought, really.
3. Duplicate the whole group and place these around to create the cluster size you are after roughly.
4. Press the Spacebar to open the search window and look for Randomize Transform. I usually choose a small random amount in Z, like 0.01, set the Z rotation to 180 degrees, tick Scale Even, and set a value of 1.10–1.20.

   This will get us into a good spot with the cluster, and it should already look quite convincing, we will then make some hand edits to get it closer to the reference.
5. I orbited around the mesh and made sure all the sides slightly leaned outward. If you want, you can add a Simple Deform Modifier in this step as well to get a bit of a bend going. Again, refer to your reference to see if this makes sense.
6. Lastly, I duplicated the whole group, rotated it 90 degrees, and pushed it down a bit to create two layers of grass, which will greatly help with how it behaves in engine. I also ran the randomize transform again to randomize the duplicate I made.

You can play around with these a little, and I recommend you do so. Compare to your reference and see what you can spot and integrate. The steps above are just a suggestion. You are in a modeling tool, so the options are limitless. Have fun with it and see if you can add your own twist.

I ended up making four clusters that you can see in Figure 3.35.

FIGURE 3.35   The blender viewport showing the four clusters I created.

To export these cleanly and efficiently, it is best to combine them in the 3D modeling tool.

First, duplicate them all so you always have your previous step available in case you need to make any changes. Then select all the meshes of a single cluster, select an active mesh and use the shortcut Ctrl+J to combine them, move the pivot to the center of the mesh, and snap to the bottom.

It is beneficial to put the pivot slightly higher in the mesh if you place it exactly at the bottom it will be less comfortable to place these around in the engine, to prevent as much floating as possible, have some of the mesh stick into the floor, refer to Figure 3.36 anything under the white line will be under the ground in Unreal, that means if the terrain slopes up or down slightly or we decide to rotate the mesh it will still touch the bottom, seen in the example on the right.

FIGURE 3.36   Two grass clusters, one it a neutral position and one slightly rotated to showcase the benefit of having the pivot slightly above the lowest point in the mesh.

Give them a name that makes sense to you; I have called them SM_Grass_01_A/ B/C/D, SM for Static Mesh, the 01 refers to the atlas; if I have multiple textures for the variations, I will call them 01,02, and so forth, and then the ABCD is to differentiate between the variations. I have chosen not to use a suffix for the LOD0 meshes as this will import into Unreal, but my LODs I have called SM_Grass_01_LOD1/2/3.

Before exporting, set them to the world origin(0,0,0) so they do not import into the wrong location in Unreal Engine later, select all meshes and use shortcut N to bring up the options panel; if you want to edit multiple values at once here, you need to hold Alt then Click the value you want to change, or Click and Drag over the XYZ Location to mark them all and then set all of them to 0 then select them one by one and export them with the same name you have given them in Blender I have placed mine here: ..*CreatingHighQualityVegetation\workfiles\blender\export*.

You can also use Blender Add-ons to help you export multiple files; I use "Blender Super Batch Export," which can be found on GitHub.

You should now have about four clusters with four LODs each. Feel free to use the kit to build more grass clusters if you want to get some more practice or if it is required for your specific use case. Refer to Figure 3.37 to get an idea of what I am exporting.

FIGURE 3.37   Four grass clusters with four LODs lined up ready for export.

That is a wrap for the Grass Clusters. The rest of the chapter will focus on creating the Wood Anemone, and we will import both of them into Unreal in Chapter 7.

CREATING THE WOOD ANEMONE

The Wood Anemone will be more complex. We will need to build multiple elements and combine them convincingly. However, the techniques will be the exact same as those used for the grass, just applied to a different subject.

When we look at the references, we can identify a couple of elements: the stem, the leaves, and the flower. It seems to be a whorled phyllotaxy where all leaves sprout from the same location. Refer to Figure 3.38 to see what I consider to be separate elements.

FIGURE 3.38   A wood anemone reference with outlines on different elements.

I consider the flower the most complex subject so let's start with building that.

Similar to the grass, we will be using modifiers to edit the geometry, but we will polish this up using some mesh edits as well; for the petals, we will not use the Curve modifier but just two Simple Deforms, I set both of them to Bend, one on the *Z* axis to control the fold and the other on the *X*-axis to control the actual bend. You can see this setup in Figure 3.39.

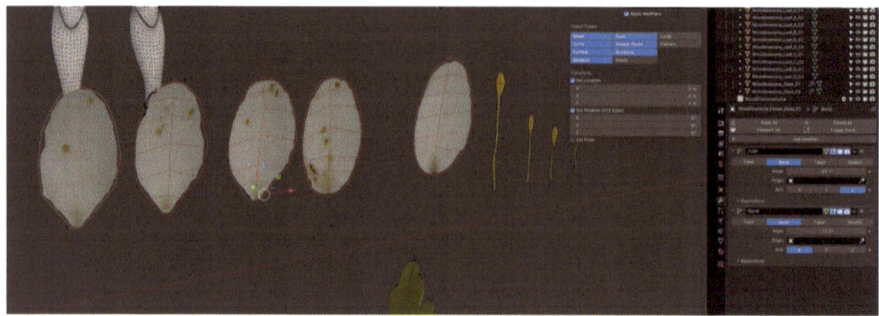

FIGURE 3.39   The wood anemone flower meshes with the modifiers visible.

With this setup, we can go ahead and start assembling the pieces, make sure to look at references while doing this.

Refer to Figure 3.40 for a step-by-step process on how I approached building the flower. Keep in mind that these flowers will be very small, so there is no need to go crazy with the amount of detail. Keep it loose, and make sure the edits you make will read well from a distance. The best way to check is to rotate around your model and zoom in and out a lot to make sure it all behaves the way you want it to behave.

FIGURE 3.40 A step-by-step process on how to build the wood anemone flower.

1. Bring in the flower bud we created earlier in this chapter, or grab it from the Blender reference file, using just the move and rotate tools to get most of it in its general place; notice how the flower petals alternate between height; there are three stacked above and three of them below. After they are generally in place, use the Bend and Fold modifiers we added earlier to add some visual interest, ensure there are no meshes clipping trough each other, and don't shy away from moving them around as much as needed to get the result you are after.

2. When using just the modifiers, you end up with a very mathematically correct-looking flower, but in real life and our references, you can see they are slightly noisy or wobbly, conveying the idea that the flower petals are soft and fragile. To add this you can select the flower petals and go to Edit Mode using Tab, then grab some vertices and move these up or down, you should see the changes happening live. In this step, it is important every once in a while to zoom out to see if the changes are aggressive enough to be noticed from a fair distance, but if you go too aggressive you will start seeing harsh polygonal shapes, so there is a balance to be struck here.

3. The last step is to get the Stamens in place; this is the pollinating part of the flower. The top is called the Anther, and the string is called the Filament; I have used the same method as the grass by adding a curve modifier. It gives me very precise control over how to place these; start by putting a couple, like three or five, then duplicate these around the flower, editing the direction as you go along; for this step, also make sure to zoom out a bit to see it from a fair distance. Make sure everything looks nice and accurate from every viewing distance.

Since we will repeat the flower quite a bit, it is a good idea to make a copy and change some things around on the duplicate. This will introduce some variation and give it a more natural look when repeated across the scene.

As seen in the reference, the flower has a stem attached to it that connects to the "main stem" later. In order to save on some elements, I have not extracted a stem specific to the flower and just reused the stem that I used for the main stem. Refer to the left side of Figure 3.41 for an idea of what the flower should look like. You can then go ahead and apply all modifiers and use Ctrl+J to Join them all together.

FIGURE 3.41   An image showcasing the isolated flower, and the first step of adding leaves on the right.

Also visible in Figure 3.41 is the next step to get the Wood Anemone strands, the main stem, and the leaves. I have placed the flower we just created on top of the main stem.

For the leaves, I added a simple deformer for the fold, and then you can choose to either add a Curve modifier or another Simple Deform. I opted for a curve to have better control. While checking with your reference, reconstruct the leaves. Pay attention to where they connect on the stem. They all connect in the same area where the flower's stem and the main stem connect.

The way I cut up the meshes for the leaves requires you to reconstruct the whole leave again, this is a bit more work in the meshing step, but it will give us better depth on the final asset, Once you have a leave reconstructed, you can go ahead and duplicate it around the stem until you get a nice round "hat" refer to Figure 3.42 for the final result.

Setting this up, I used identical techniques to the grass and flower, but if you prefer another method or like to play around with it yourself, feel free to do so. The important thing is that it looks like our reference. The road you take to get there is entirely up to you. I do recommend pursuing non-destructive workflows, though.

This concludes a brief overview of how to build each element, and all that remains is creating a couple of variations, when making variations, it is important to not make them too similar to each other, this way, each variation brings something to the table, so vary them in height, width, the way they bend, where the leaves come off, but also think about the silhouette try to find as many opportunities you can to bring in variations. The source files include four different leaves; two of them are larger and are used mostly on the larger strands, and two of the leaves are a bit smaller and differ in shape; these are used for the lower variations.

FIGURE 3.42   The wood anemone with the flower and leaves combined.

I ended up building six different strands, as seen in Figure 3.43. With these six clusters in our toolbox, we will create a cluster in the same way as we did the grass, as a quick recap:

FIGURE 3.43   Six variations of wood anemone flowers.

Firstly, place all strands in cluster, then duplicate this cluster around a couple of times until you reach a good size for your cluster, bring up the search bar and search for Randomize Transforms, and use this to quickly create variation, once you have this setup you can go in and make some deliberate changes, notice on the reference that the whole cluster grows in a dome-like shape, so make sure to rotate some of the ones on the side outwards.

Additionally, for the Wood Anemone, we need to decide how many flowers we want per cluster and where to place them; I placed mine primarily in the center so we can use the outer edges to blend the pieces a bit in Unreal Engine.

Refer to Figure 3.44 for the variations I built for the Wood Anemone. You can build as many variations as you like, but keep in mind that they should all contribute something; try to avoid making variations that look too similar.

FIGURE 3.44   Six variations of wood anemone clusters, ready to be exported.

1. Is a close-up of the first variation.
2. A lineup of all five variations. They differ in height, and I included some clusters that do not contain any flowers at all.

If you intend to build LODs, now would be a good time to do so. If not, we can move on to the next chapter; make sure to name and export your Wood Anemone clusters at 0,0,0 as well; I called mine SM_WoodAnemone_01_A/B/C/D.

Setting up in Unreal Engine

4

SETUP IN UNREAL ENGINE 5

With our first assets ready to be imported let's open Unreal Engine 5, you can do this by downloading and starting the Epic Games Launcher here https://store.epicgames.com/en-US/download.

With the launcher open, navigate to "Library" and add an engine version by pressing the + icon at the top of the launcher and start the engine.

I have used Engine Version 5.3.2 which gets continuously updated and there will be newer versions available; there is always a chance that things break or stop working as Unreal Engine updates so if you download a newer version, it could be that the source project packaged with the book will no longer work as expected, with that being said, I do always recommend working on the latest version of the editor so if you find any issues with newer versions, I suggest seeing if you can figure out a work around or fix it.

If you are unable to do so, there is always the possibility to download older versions of the engine.

The first time you open the engine, it will ask you to create a project. The process is described in Figure 4.1:

1. First, you will have to click on "Games"
2. This is where you select the type of project you want. In this case, I selected the Third Person. This will give us a third-person template, allowing us to walk around in it with an already set-up player character. This helps to judge the scale at a later stage.
3. At Project Defaults, you decide on some specifications for the project. This can be adjusted later, but in this instance, you can keep it on Blueprints. The target platform is Desktop, and we will use the Maximum quality pre-set and turn on Raytracing.
4. The project location will be a location on your disk if you are opening up the project that comes with the book; you can find it here: ..\CRC\ *CreatingHighQualityVegetation\workfiles\ue*

   I recommend saving it in a place that makes sense for you, it will be difficult to change this at a later stage.

DOI: 10.1201/9781003492283-4

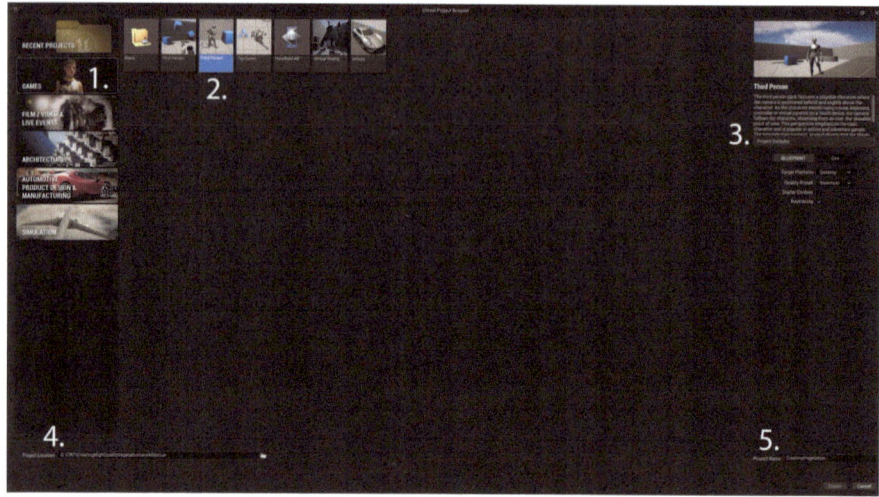

FIGURE 4.1   The Unreal project browser.

5. Lastly, you can define your project name here. I have called it "CreatingVegetation" but feel free to name it as you want.

When you open the project, you should see the default map that comes with the project type you have selected in step 2 on Figure 4.1, and we can start to build our folder structure; you do this in the "Content Browser" you can open the Content Browser by pressing Ctrl+Space. This will open a tab in whichever window of Unreal Engine you are in. Press "Dock in layout" in the top right corner if you want it to remain open (Figure 4.2).

FIGURE 4.2   The content browser.

The content browser shows you all the files that are in your project.

1. This shows your folder structure. This is mirrored to what you have on disk.
2. Shows the actual content inside of that folder.

If you hover your mouse over the Content folder, Right-Click, and say New Folder, a new folder will be created that we will use to store all of our data. I called mine "CreatingVegetation" and within that folder I created three other folders called Assets, Levels, and Materials. For now, that will be enough.

I find it easiest to work in a level created by myself, so go to File, New Level, or press Ctrl + N and create a Basic level; this will give you a scene with a basic light setup, a skybox, and a Floor in the center of the map. For now, we can keep the Floor there to have something to walk on, but later on, we will remove this. Go to File, Save, or press Ctrl + S and save this level in the recently created Levels folder; I called my level "Main."

To make this level the default level and open it on start-up, go to Edit and select Project Settings. Click on Maps & Modes in the left panel to open that subsection of Project Settings. Click on the dropdown arrow to the right of the Game Default Map field, then select your desired Level from the list. Or alternatively Search for "Default Map" at the top of this window and make Main your Editor Start up Map and your Game Default map.

With our initial setup done, let's import our grass assets and set up a simple material.

Go to your Asset folder and make a folder called Vegetation and a folder within that is called "Grass" Right-Click in that folder and select "Import to" you can refer to Figure 4.3 for an example.

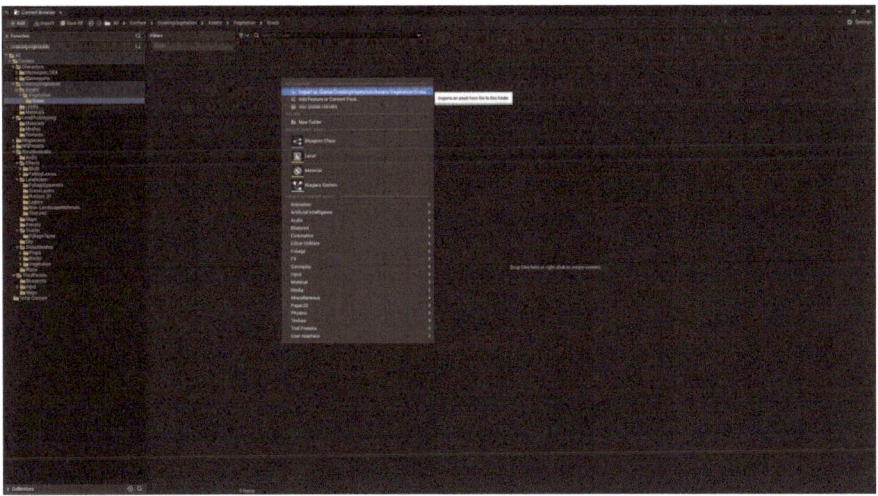

FIGURE 4.3 Importing a new asset into the content browser.

IMPORTING OUR ASSETS AND SETTINGS UP LODs

We can then go ahead and import the grass that we exported from Blender before; I exported mine here ..*CreatingHighQualityVegetation\workfiles\blender\export*. Make sure to import LOD0. We will set up our LODs later in this chapter, but for now, we can leave them alone.

You will be presented with the FBX Import Options menu on your first import. This menu is a chance to change some preferred options when importing meshes. There are many settings here, but the ones we are interested in are under the Mesh and Material Tab. In the Mesh Tab, we want to change the Vertex Color Import Option from Ignore to Replace if you hover over these settings, it will tell you what they do, but Unreal does not automatically import custom vertex colors, since we will be needing those for our Wind Setup in Chapter 10 we need to set this to Replace instead. Then, under the Material tab, set the Material Import Method to Do Not Create Material and untick Import textures.

This will ensure we only import the mesh data and not have Unreal create additional data. I find this a cleaner way to work and allows me to control what goes where without risking bloating the folder structure.

You should now see a thumbnail of your asset that is named identically to the FBX you imported. You can Right-Click it to rename it to something else or use shortcut F2 with the selected asset. I advise doing this if you haven't set a name for it before.

If you double-click the asset, it will open up the Asset Viewer; here, we can change and add any data that is specific to this asset, such as collision, materials, and LODs; if you have built LODs for the asset, now would be a good time to import those. On the right side, you will find a Tab called LOD settings; under LOD import, click the drop-down menu and say Import LOD Level 1. Click that and navigate to your LOD1 mesh; once that is imported, you can repeat this step for LOD2 and LOD3. Refer to Figure 4.4 to see where this is located.

If you want to see what happens more clearly, you can go to the top left of the viewport and click Lit, then Level of Detail Coloration, and then Mesh LOD Coloration to change the display to a debug display specifically for LODs. If you move the camera closer or further away, you should be able to see the colors switch. This is the mesh switching to a different LOD. This is also reflected in the data window in the top left; the triangle count should be lower when you move the camera away.

The moment Unreal switches is based on screen size. By default, this gets an automatic value, which I usually think is a bit too harsh. And I find it switches to a lower quality relatively quickly. Fortunately, we can set these values by hand.

To do this, go to the LOD Picker Tab and tick on Custom; this will give you new Tab called LOD0 to 3 or however many LODs you have imported, and we can set the Screen Size manually – I recommend keeping the LOD coloration on while changing these settings.

FIGURE 4.4   The asset viewer with the LOD settings highlighter.

These settings depend a bit on your object size, but you can see if the numbers I used work for you:

LOD0 Screen Size: 1.00
LOD1 Screen Size: 0.4
LOD2 Screen Size: 0.2
LOD3 Screen Size: 0.15

Notice how the gaps get smaller the further down I go down the LOD chain; this is because once the object gets further away, it is okay to switch a bit harsher, but I would like to keep everything close to the camera at the best possible quality. For personal scenes, LODs are not that important, but they are incredibly important in real-time environments constrained by performance; for all AAA games, a great deal of time is spent making LODs as good as possible and ensuring the transition between LODs is nearly invisible.

That concludes setting up our asset if you have created the Wood Anemone or any other example plants, you can repeat the same steps for those before making the material.

CREATING A BASIC VEGETATION MATERIAL

To get a textured view of our assets, we will need to create a material; at a high level, you can see the material as the "paint" applied to your object, but in reality, it is more complicated than that, in abstract terms a Material is used to calculate how light interacts with the asset using various textures and/or math expressions.

It sounds more complicated than it is, so let's start with creating a straightforward material to get started. In the Materials folder, Right-Click and click on Material. I have called this M_VegetationMaster. M_ stands for material and is part of the Recommended Unreal Engine naming convention. We will be using more of these prefixes so to familiarize yourself with them you can check out this page of the Unreal documentation https://docs.unrealengine.com/4.27/en-US/ProductionPipelines/AssetNaming/.

Once you have your material created, double-click it to open up the material editor, which should look like Figure 4.5.

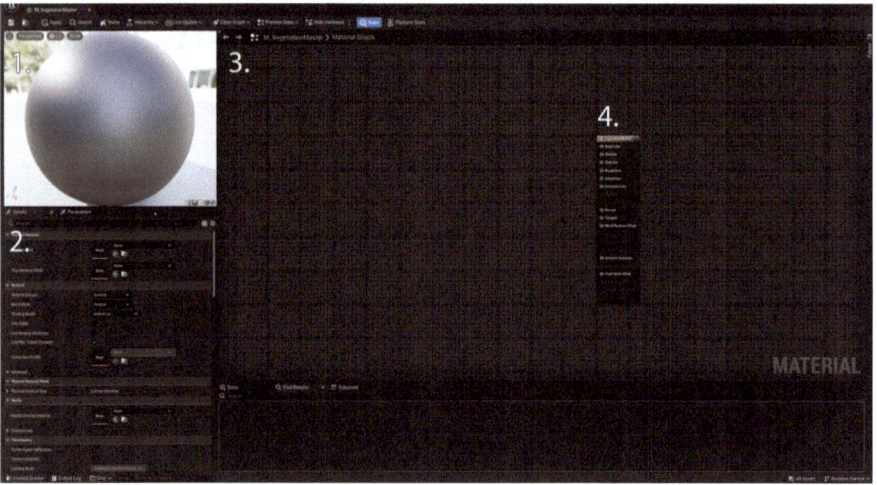

FIGURE 4.5 Overview of the material editor.

We will be looking at a couple of things while in the material editor.

1. This ball shows you a preview of what your material is going to look like.
2. The details pane contains all the data you can edit for each node.
3. This is the material graph, the editor is node based and any new logic is created here.
4. The node that is already there is the main node, this contains a lot of important information on what type of material it is, and will be the first node we will have a look at.

With the main node selected, there are three things that we need to change in the details panel. Starting from the top, the Blend mode needs to be set to Masked, enabling us to use our Opacity Mask. Set the Shading Model to Two-Sided Foliage; this will allow light to pass through the surface using Subsurface scattering, which is necessary to accurately represent leaves; we will mask this out on things like veins and branches. To ensure we see our Material from both sides of the surface, we also need to turn on Two Sided.

To be able to load in our textures you can do two things, the simplest way is to drag in your texture into the node editor and plug it directly into the corresponding slot, but the smart and efficient way would be to prepare this material for Material Instances.

Material instancing is a way to create a parent Material that you can use as a base to make a wide variety of different-looking children (Material instances). To achieve this flexibility, Material instancing uses inheritance: the properties of the parent material are passed to its children.

The other benefit of using Material instances is that if you are working on larger scenes, it is cheaper for the computer to process instances than it is to process a master material for each individual asset, giving us an increase in performance and a decrease in workload.

For our textures to be editable in the Material Instance, we need to add a Texture Sample; you can do this by right-clicking on the grid and searching for "TextureSample." After that you can Right-Click the Texture Sample and click Convert to Parameter. For now, we will create three parameters Base Color, RoughnessSubsurfaceMask and Normal.

The TextureParameter has multiple outputs RGB which is the combined channels excluding the alpha channel. R, G, B, and A allow us to separate the texture into its different channels, and RGBA, which is the complete texture and includes the alpha channel.

For the vast majority, only the top four are relevant. RGBA will output the texture exactly as is with all channels intact; however, since we packed different types of data in our texture, we need to match that with our set up in engine, in our case for the Base Color we can plug RGB into Base Color and A, which stands for Alpha Channel into the Opacity Mask Input.

Refer to Figure 4.6 for an example of this setup.

For the RoughnessSubsurfaceMask we use the R channel and plug that into Roughness for now. Also visible in Figure 4.6.

You will notice that you get an error on all three parameters, if you have a look at the stats window at the bottom, it will say *[SM6] (Node TextureSampleParameter2D) Param2D> Found NULL, requires Texture2D,* which is essentially telling us that all these are looking for a texture file, I like to keep this clean and not load in a texture file from an asset but use a set of debug textures instead, you can find the ones I use in: ..\CRC\CreatingHighQualityVegetation\workfiles\ue\debug.

You will find two textures there, T_Gray_C and T_Normal_C I have imported these into a Debug folder inside the Vegetation folder and applied them to the Texture parameters.

Your result should look something like Figure 4.6 and there is one more thing we can do to keep this organized if you look at point 2 on the same figure. Currently, the Group is set to None; as we work further on the material, we are going to want to group different parameters in different groups. We will get into this shortly, but for now, you can type Textures in that field and then use the dropdown menu in the other parameters to assign them to the same group.

To get a better idea of where this will end up, we can go ahead and create a material instance and assign this to all our grass meshes. To create a Material Instance, Right-Click the M_VegetationMaster Material and say Create Material Instance at the

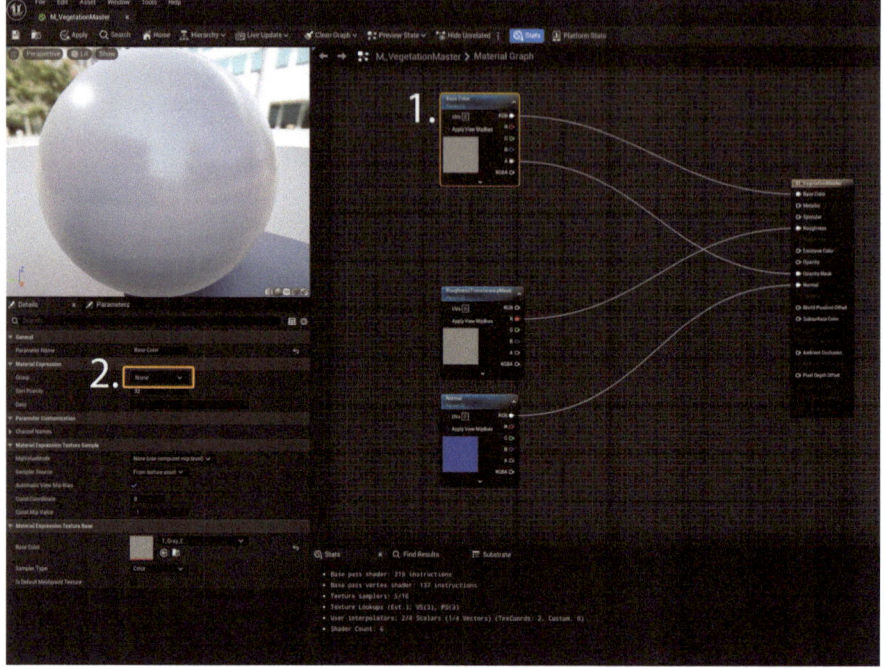

FIGURE 4.6 The material editor with three texture parameters visible.

top of the menu. I called mine MI_Grass_01; in this case, MI_ stands for Material Instance. By default, you will notice that Unreal will place the Material Instance in the same folder as the Master Material, so do not forget to move your instance over to your Grass folder you can do this by simply dragging the Material Instance into the folder you want it to be, it will give you the option to say Copy here or Move here. Select Move.

If you double-click the material instance, it should open the editor seen in Figure 4.7 highlighted in orange you can see that the parameters we have created have ended up exposed in the Textures group, by default these are ticked off but in the figure I have gone ahead and turned those one.

With these parameters exposed, we can now navigate to our grass textures in the content browser and drag them into the corresponding parameters of the material instance. Alternatively, you can browse to the texture in the content browser, select it, and hit the arrow under the parameter name that states "Use Selected Asset From Content Browser."

With our basic texture applied, it is a good idea to apply this to our grass meshes and see what it looks like in the engine. Go ahead and open up the mesh editor for our grass, you can do this by double-clicking on the grass meshes in the content browser and applying the material instance to all our variations; you can either search for the

FIGURE 4.7   The material editor with three texture parameters visible.

Material Instance created for the grass in the top right of the details panel (inside the Static Mesh Editor) or by dragging the Material Instance directly from the Content Browser onto the Material Slot.

Some additional parameters that we can add to enhance flexibility and allow us to do the finishing touches in engine are Base Color Brightness, Specular, Roughness Intensity, Normal Intensity, and last but not least Subsurface Strength and Intensity.

All of these can be achieved using single values and simple math. Open up your M_VegetationMaster material and start with the Roughness and Specular intensity. By holding down the 1 key on your keyboard and clicking on the grid, you will create a Constant1. Alternatively, you can Right Click on the grid and search for Constant. Similar to the textures, we need to Right-Click this node and say Convert to Parameter, create two of these, and call them SpecularIntensity with a default value of 0.5 and RoughnessIntensity with a default value of 1.0. For Group, we can create another group called "Adjustments."

The SpecularIntensity can be plugged straight into the Specular input, and for the Roughness, we need to create a Multiply node, hold down M on your keyboard and click on the grid, or Right-Click and search for Multiply to create one. Then plug the R channel of the Roughness Texture Parameter into slot A and the RoughnessIntensity Constant1 into the B slot, then plug the multiply node into the Roughness input instead of directly from the TextureParameter.

This can sound a bit complicated in text but you can refer to Figure 4.9 for an example of what we will be setting up.

Don't forget to save, and let's move on to Normal Intensity.

Normal Intensity requires more logic and understanding of how a normal map works.

In a nutshell, the RGB values of a normal map correspond to the XYZ vectors: R decides the direction on the X axis, G on the Y axis, and B controls the Z axis. Z is used for depth, so if we edit all channels at once and lower the Normal map intensity, it would be the equivalent of pushing the Z axis in; this results in a darker and incorrect shading, so step one is to split up the RG and B of the normal map. This can be achieved with a ComponentMask node, so create two. For the first one, select R and G; for the second one, select only B.

We want to leave the normal maps' blue channel alone and edit only the red and green channels, so put down a Multiply node by holding M and clicking on the grid, connect the RG mask to the Multiply A slot, then add a Constant1 parameter with a default value of 1 into the B slot of the Multiply node. Call this Normal Intensity and add it to the Adjustments group.

After that, we need a way to bring back the RG and the B channel so it is seen as a Vector3 and is accepted by the input of the main node. We can do this using an Append node, as per usual, Right-Click inside the grid and search for AppendVector, plug the Mask(RG) into A and the Mask (B), and plug the whole logic back into the Normal input. Your final result should look like Figure 4.8.

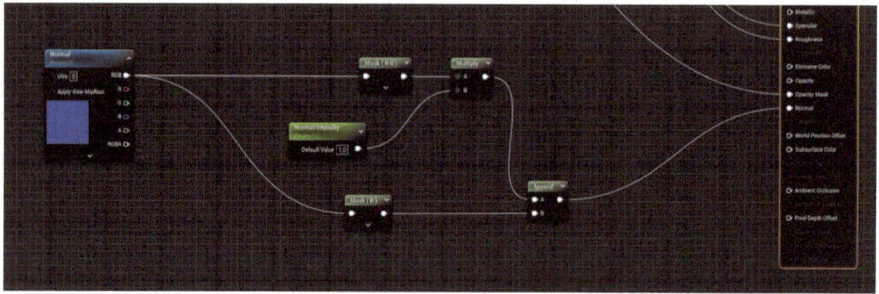

FIGURE 4.8 Viewing the material editor with the setup needed to control normal intensity.

Lastly, we are going to look at the Base Color and Subsurface intensity. For this, we will not need any new nodes or knowledge, and we can use the techniques we just learned. Let's start by putting down another Multiply and Constant1 parameter with a default value of 1 called BaseColorIntensity; plug the Base Color into the A slot and the parameter in B, then plug this back into the Base Color Slot.

Repeat the setup above, except this time, we call the parameter "Subsurface Intensity" and we need to ensure that this gets the data from our earlier created Subsurface Mask. To do that, put down another Multiply node, plug our previously created Multiply node into the A slot, and multiply this by the green channel of our RoughnessTranslucencyMask.

This wraps up most of the texture logic for now, and we can start setting up our material instance to see if everything works as intended. The final logic should look like Figure 4.9.

If everything works as intended on the grass, you can create a new Material Instance for the Wood Anemone and apply it to that as well.

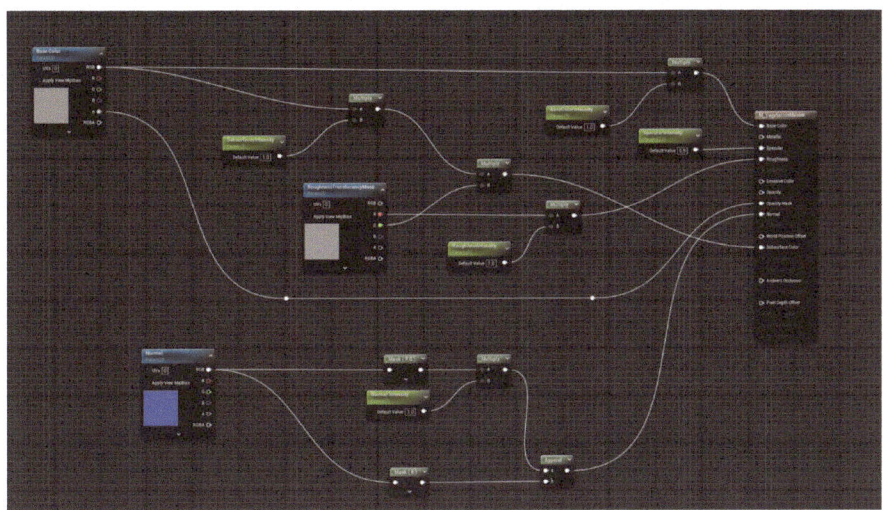

FIGURE 4.9   Viewing the material editor with all the editable parameters assigned and visible.

Photogrammetry Explained

5

WHAT IS PHOTOGRAMMETRY?

Photogrammetry is the act of obtaining information about digital objects or environments through the interpretation of photographic data and patterns. It is not new or exclusive to the video games industry but has found many use cases there. Essentially, it means taking a lot of images of a single subject, in our case, tree trunks, barks, and other elements of a tree, and having a computer process these.

A quick disclaimer: the files needed and generated through photogrammetry are very large and for that reason they are difficult to share, combined with the fact that the processing happens for the majority automatically with the exception of the baking and optimizing of the in-game meshes.

If you do decide to go out and do your own scanning, you can use the book and additional online resources as a guide, but if you intend to use the scanned trunks provided, this is a read-only chapter.

I will be using RealityCapture to process photogrammetry data. If you want to try photogrammetry for yourself, there is an abundance of resources available on their website. If you do not have a good camera, you can download the RealityCapture app for iPhone and Android and follow the on-screen prompts; doing this yourself will teach you much more than any book or video ever can, so I highly recommend you to give it a shot before moving on.

For those who are looking for the files created in this chapter, you can find the processed and optimized trunks here: *..\CRC\CreatingHighQualityVegetation\workfiles\blender\export\Phototrunks*

Inside RealityCapture, we will process our images, create a point cloud, and convert the point cloud into a high-resolution mesh that we can use and manipulate to create game-ready content.

Since photogrammetry data will give us data based on real images, it is essential to keep some things in mind when trying to find your subject. As a general note, it is best to scan on clouded days. If you scan while the sun is out you will capture all that lighting data and it will be nearly impossible to create a neutral texture from that.

DOI: 10.1201/9781003492283-5

CAPTURING TREE TRUNKS

When capturing tree trunks, try to find relatively straight trunks; that way, you can do all the deformations in the mesh later. Additionally, trees with many roots usually look majestic and are exciting topics to recreate digitally. However, they do not lend themselves well to game environments as they stand out too much from the crowd; finding something generic looking trees will allow for much more repetition and will give you an easier time building a scene.

If you find a tree trunk with many unique features, both large and small, it is probably best to avoid it if you want to repeat the asset. In most cases, things like burls, scars, and repetitive lichen growth are unwanted.

Of course, when creating a hero asset that has no intention of being repeated many times, it is okay to go for more unique features.

For this reason, if you go out scanning, I would recommend going to a forest where you have a large selection of trees and scanning a large variety while you are at the location; there is nothing worse than coming home, reviewing the data, and figuring out you missed a couple of things, I usually try to capture different ages as well, I find saplings working very well for smaller trees but also to extract tiling textures from to be used in second generation branches. We will look at how this is done later in the chapter.

Additionally, it is better to be safe than sorry when taking photographs. If you miss something while taking photos, RealityCapture will not know what to do with it, and it will look like the data is broken; there are ways to fix this by, for example, cloning other parts of the scans and placing them in the missing areas but is much easier and better looking if all the data is there and can be processed so make sure to capture all the nooks and crannies to avoid having to fill in any gaps later.

After you take your photos, you can process them using RealityCapture. I usually feed RealityCapture two sets of data: one for the meshing, which has a slightly higher contrast, and one for the texturing, which has some of the blacks and shadows already softened to help with delighting later. This is explained in detail in RealityCaptures introduction YouTube videos so I recommend checking them out if you would like to see more.

Refer to Figure 5.1 for a brief explanation of what happens in RealityCapture.

FIGURE 5.1 Different stages of photogrammetry processing.

1. The dots you see floating are the location of where the pictures have been taken, as you can see, I circled the subject and took images at different heights, ensuring there is some overlap in each picture; RealityCapture will align these cameras and from there a Point Cloud is generated, a point cloud is created by finding similar points in the photo, and by interpreting the overlap RealityCapture defines a location in 3D space. Think of a point cloud as set of data containing coordinates which will be used as a base for the meshing later.
2. The meshing step is where all the photos and point cloud get further analyzed to create a high-density mesh.
3. The high-density mesh then gets automatically unwrapped and textured using the photo data, as you can see on the image, even though I shot this on a cloudy day in the shade, there is still a lot of shadow and lighting data left, this will be further removed when delighting the mesh later.

With the high-fidelity data collected, we will need a low-poly mesh for this and rebake the texture onto it before we can comfortably delight our texture.

REMESHING AND REBAKING THE HIGH-FIDELITY DATA

Creating a low-poly mesh is the same across all forms of modeling. If you have never done this before, I recommend watching some retopology videos. By far, the quickest method would be some form of automatic decimation, but this results in a very messy topology. You can refer to Figure 5.2 for an example of a decimated mesh. This was

FIGURE 5.2 A decimated mesh with a lot of mesh errors present.

done in a couple of seconds, but the clean-up will take much longer than that, and even after the clean-up, there are some negative side effects to consider.

To avoid having to use decimation, I have used a combination of Zbrush's Zremesher tool to quickly create some topology that has quads (square shaped geometry), then took this into Blender and used the PolyBuild tools to adjust and create new geometry to clean up the automated result I got from Zbrush. A quad mesh has a couple of benefits:

1. A quad mesh will subdivide much better if we use some form of displacement later. As you can imagine, splitting a square in two is much easier than splitting a triangle in two. Additionally, it decreases the risk of lighting errors due to faulty geometry.
2. When unwrapping this mesh later, it will be much cleaner and easier to work with the UVs; decimation tools create very jagged geometry, resulting in jagged UV seams. They are hard or impossible to straighten and smooth.
3. Quads are more predictable and easier to manipulate when working on the mesh.
4. It is not as important when working on a project yourself, but if you join a company and find yourself in a situation where your meshes need to be shared, having a proper topology will make it much easier for the next person to do their work.

It all boils down to the same thing: I described this before when talking about destructive and non-destructive workflows: once a decimation has been done, the workflow has become destructive and no longer lends itself to comfortable editing; I recommend only decimating as a last resort.

As a finishing touch, I defined a ground plane and used Blenders Boolean tools to remove anything underneath the ground plane. This will save triangles but also allow me to pack the UVs more tightly, resulting in a slightly higher texel density.

Refer to Figure 5.3 to see the high-poly and low-poly mesh side by side.

As you can see, the high-poly mesh in Figure 5.3 is so dense that the wireframe turns completely black; refer to the zoomed-in part to see the density of triangles. On the right side, we see the retopologized tree trunk; a UV checker texture is also applied.

A UV checker is a texture containing straight lines and numbers, which allows us to efficiently check for errors in the UVs. In Figure 5.4, you can see the difference. The left side is a clean, properly laid-out UV, and on the right side, you see warping and stretching.

Using a clean texture with numbers, letters, and/or straight lines allows us to differentiate between a UV error and something inside the texture. If we had applied a bark texture, there would be no telling if it is warped because of how the bark grew or if it is happening because of a technical error in the UVs.

The high-poly and low-poly were then taken to Substance Designer; inside Substance Designer, I baked the mesh data and transferred the color map. With the color map Transferred, I also removed a lot of lighting data to delight the texture; I used a combination of the Ambient Occlusion map and the normal to target specific areas.

Again, because I scanned this on a cloudy day, all the lighting data that was there was essentially ambient occlusion; once I baked my ambient occlusion, I could use that to counter the baked-in Ambient Occlusion.

FIGURE 5.3   The high-fidelity mesh and low-poly side by side.

FIGURE 5.4   UV checking texture.

Refer to Figure 5.5 to see the side-by-side. The right mesh is de-lit. It might look a bit bland in the viewport, but once we get this in the engine and start lighting it ourselves, a lot of that information will come back, but instead of being baked in a texture, it will work regardless of the light angle.

FIGURE 5.5   Side by side of the original trunk and the trunk with its lighting data removed.

Usually, when delighting, you run a risk of removing more data than you get back, so it is a good idea, once you have your trunk or tree setup in Unreal, to compare the two and bring back any shadows you have lost and are not being regenerated in the engine.

CAPTURING TILING TEXTURES

We will also need a couple of tiling textures on top of the trunk. A tiling texture is a texture that repeats and wraps around itself. This is the most flexible way to create branches and longer trunks.

Because the texture repeats, you want to have as few unique elements as possible. Editing the texture by hand to remove repeating elements will most likely be required. For this reason, it is very important that if you are scanning a trunk, you try to find one that is as clean as possible. No lichen, no moss, no burls, cuts, or scars, but also think about discolorations; I find it useful to scan a bit higher up on the trunk to avoid as much discoloration from the ground as possible.

When texturing, in general, it is good to focus on details from every distance; if it is seen from far away, you want to see some macro variation, slight discoloration, and larger shapes, often referred to as Primary shapes, but when you get closer, you want to start discovering some details, some lichen, bug bites or things that suggest a life cycle, for leaves this can be discoloration, dried of dead leaves or just a subtle color variation on each leaf.

Striking a good balance in Primary, Secondary, and sometimes even Tertiary shapes will be the difference between an okay and a great texture, so always look for opportunities to include them.

In my case, I found an easily accessible Beech trunk without too many unique details, and I captured about 1.5 meters of it, My intention is to create a cylinder of 1 meter so I know how long my texture is physically, so if I would end up building a tree that is 15 meters, I know I need to tile my texture about 14 times (15 meter minus the trunk) for a physically accurate representation.

This was processed identically to the trunk and then taken into Blender; I used a Shrink Wrap modifier to wrap a cylinder around the high-poly trunk and unwrapped this; refer to Figure 5.6 to see an example.

FIGURE 5.6   The low- and high-poly for the tiling texture, with the unwrap on the left.

Notice that the unwrap is aligned at the outer edges, otherwise referred to as the 0 to 1 space within the UVs; this ensures the entire circumference of the high-poly will be baked down, and we automatically tile the texture over the horizontal axis. All that is left is to make the vertical axis tiling. We can do this in Substance Designer.

I have used the Make it Tile Photo Color Node to get the vertical axis to tile and used a couple of other nodes to remove repeating elements, like Clone Patch Color, Color Equalizer, and Color Match, if you are interested in reverse engineering this to see what happens in detail, you can refer to the substance file in the example files ..\ *CreatingHighQualityVegetation\workfiles\designer\SBS_BeechPhotogrammetry_ Processing.sbs*

You can also refer to Figure 5.7 to see a summary of this process.

FIGURE 5.7   Overview of what steps were taken to create a tiling texture.

1. This is the base texture I got right after the bake. There are some obvious repeating elements present: the large discoloration, the lichen-like growth, and some of the larger "eyes" or knots. They might not look terrible on just a single tile, but if we repeat this 15 times across the whole trunk, you can imagine how these things quickly become repetitive and noticeable; therefore, it is better that they are removed.
2. I used a couple of masks created with gradients to isolate the large discoloration, then used a Levels node to brighten these areas. For the repeating elements, I used a Clone Patch node to clone another area of the texture on top of those spots; I repeated this process for all baked maps so they are uniformly fixed across all textures (Color, Albedo, Ambient Occlusion, etc.)

   This is also the step where it is made tiling by using the Make It Tile node and targeting just the vertical Axis.
3. Even after this, you can still see a large variety of color variations, which, when tiled, risk looking noisy and repetitive, so I opted to balance these out using a Color Equalizer node. I scanned the trunk and tiling texture on the same day, but as the sun went down, the colors slightly changed in between scans, so on top of all the edits to remove the tiling elements, I also used a Color Match node and matched the colors to the colors of the scanned trunk.

With all that done, we can now move on with a nice tiling texture, I do recommend checking it out quickly in Blender to make sure you are happy with the result, preferably over a bit of a larger surface so you can judge how it tiles, refer to Figure 5.8 for

FIGURE 5.8   The original low-poly mesh and a longer cylinder to check the tiling.

an example, the left is our original mesh, and the right is where I am tiling it about three times, as you can see no obvious repeating elements, and overall it is pretty hard to notice a pattern. Which is precisely what we are after when creating tiling textures.

WHAT ELSE TO CAPTURE?

If you are capturing and are wondering what else to capture? The answer is really whatever else you see, we will not cover them in detail in this book, but if you want to take what you have learned to the next level, think about capturing moss, burls and scars, lichen growth or mushrooms, anything that will help you decorate your tree trunk with additional details or set two of the same trees apart visually is of value, don't shy away from experimenting with this and remember it is often better to capture more and use less, than the other way around.

We can now take these elements and use them as a base in SpeedTree.

Creating a Tree Procedurally

6

In this chapter, I will explain how to create a tree procedurally. In my case, it is a Beech tree, and if you are using the source files, the same will be true for you.

If at any point you get stuck, you can find my SpeedTree Files here: ..*Creating HighQualityVegetation\workfiles\speedtree*

An important thing to notice is that during the writing of this book, SpeedTree has updated to version 10; in this chapter, I have used SpeedTree 9.5.2. It is not a problem to use SpeedTree 10, but they have updated the way the UI looks, the buttons are in the same place, but they look a bit more modern in SpeedTree 10; this should not hinder you in following along with the book but, if at any point you feel stuck I recommend looking at the official SpeedTree documentation for clarification.

Everything before has been explained assuming a manual building process. The major difference is that if you are building manually, you are building results – it's a straight-forward linear process where you can copy most of your references almost one-to-one, where you are in full control of the shapes and which areas you are editing.

When building procedurally, the way of working changes completely; instead of building the result, you are building the process to get there, in a way simulating what could happen to a tree in real life, if they are affected by wind, how do the branches bend, where would the branches break, what is the length of the second-generation branches compared to the trunk you should strive to recreate these rules so you can generate an infinite amount of trees using the same rule set. But it is important to let go of wanting to control every aspect of the tree, a procedural tool is not meant to be micro-managed, and a good way to view building procedurally is that you are guiding the tool and tree in the right direction.

Many procedural tools have manual editing functionality, but as a rule of thumb, I recommend setting it up so you have around seventy to eighty percent of the tree set up procedurally and only do the finishing touches with a manual approach.

Because procedural work can be vastly different between people, depending on which settings you use, you would have to copy all my values one to one, and any hand edits I do will need to be replicated to perfection if you want the exact same result. This is not feasible, and I strongly recommend you look at the screenshots in the book as a reference on roughly where your tree should be at that given step.

Realize and accept that yours will not look the same. Take some liberties and bring in your own creativity. Compare to your main reference and build that. Instead of trying to follow along. Your focus should be on understanding what thinking and working procedurally means.

DOI: 10.1201/9781003492283-6

To create the tree procedurally, we will be using SpeedTree. You can download a learning version for free, but if you want to export your assets, you will need to purchase a license. They have flexible options and they are available here: https://store.speedtree.com/

After downloading and installing, you can go ahead and open SpeedTree. We will start by setting some preferences and then import our photogrammetry tree trunks and tiling texture. After that, we will create some nodes to create the initial meshes for our tree. Speedtree is a node-based program, meaning that individual components can be edited in their corresponding nodes.

When you first open SpeedTree you will be greeted with the window you see in Figure 6.1.

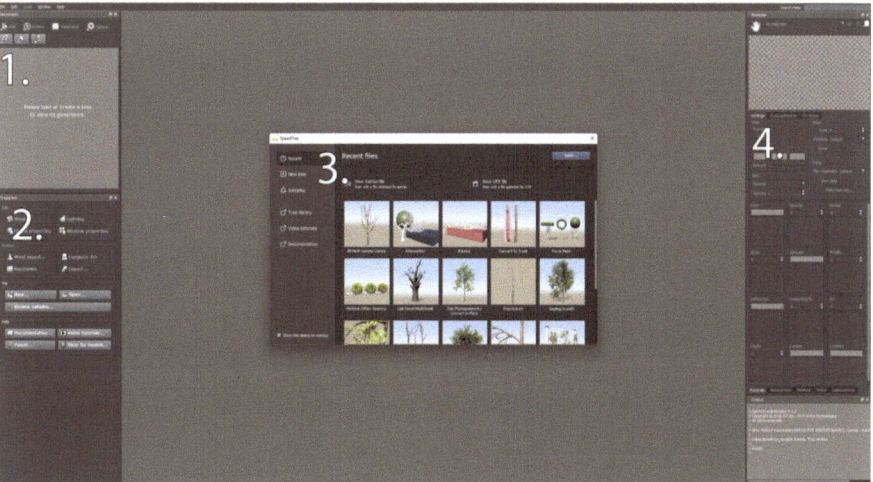

FIGURE 6.1 A screenshot showing the very basics of the SpeedTree UI.

1. In this area, we will be creating our node network. This is where you define the type of geometry you are creating, which is essentially the hierarchy of the tree we are building.
2. If you have a node selected, all its settings will be visible here.
3. This shows your recent files right now. For our purposes, we will select New Game File, which will open a new scene, but I recommend to also try opening some of the sample files and playing around with those. One of the best ways to get familiar with SpeedTree is by playing around and trying to understand these sample files. Often, I find myself copying over some of the nodes in the sample files to edit and use in my own trees. There is no need to re-invent the wheel.
4. This section of SpeedTree is used for creating and managing materials, meshes, and displacements. We will use this window quite often when creating our leaf meshes and textures.

With the New Game File selected, and a fresh SpeedTree file opened, I recommend you click around a bit, familiarize yourself with how the viewport works, Right-Click in the Node Network and create some geometry, change all the settings you can find, and see if you understand what it does – do not commit to anything just yet but just have some fun with the software.

In my opinion, the best way to learn SpeedTree is to touch every setting and see what it does; if it is not immediately clear to you what it does or what it means, try googling it or looking it up in the official SpeedTree documentation. Even though you might not use every setting for your trees, it is good to know what else is possible; filling your toolbox with knowledge about different settings might set you off in the right direction in the future.

Once you are done clicking around, Right-Click on the node graph, add Geometry, Photogrammetry, and select Mesh.

This will create a mesh node, and you should see just a cube in the viewport, refer to Figure 6.2.

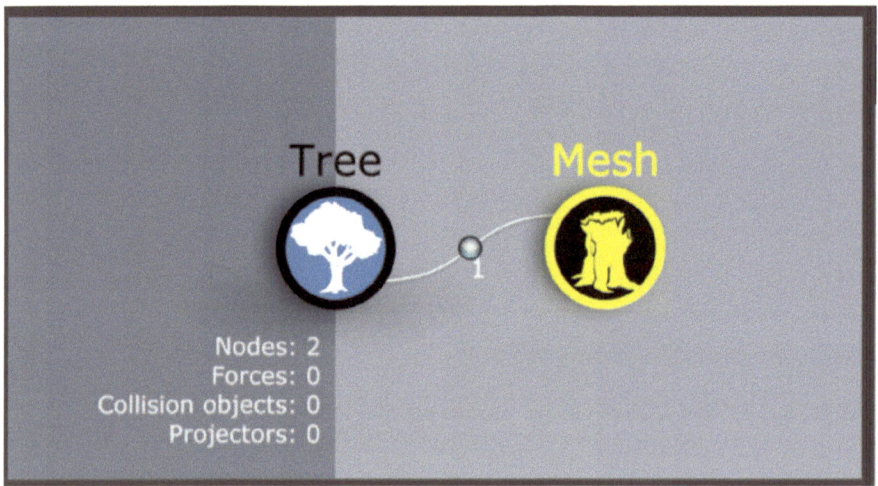

FIGURE 6.2   The node graph with a mesh node visible.

Now would be a good time to save the file, I saved mine here: ..*CreatingHigh QualityVegetation\workfiles\speedtree* and called it ST_Beech_L_01.

The concept of adding meshes and materials can be perceived as a little backward in Speedtree, as you can't directly assign a mesh to a node, but you can assign a Material. So what we need to do is create a new Material, assign a Mesh to that Material and then assign the Material to our mesh node to get both the Mesh and Material visible.

Navigate to the Materials Tab on the right, and select the +/− button on the top right, this will open up the Materials window where you can create and manage materials, Click Add New and name it Beech_01_L_Phototrunk, while we are here, you can go ahead and make a second material and call this Beech_Tiling, we will ignore that one for now.

Refer to Figure 6.3 for reference.

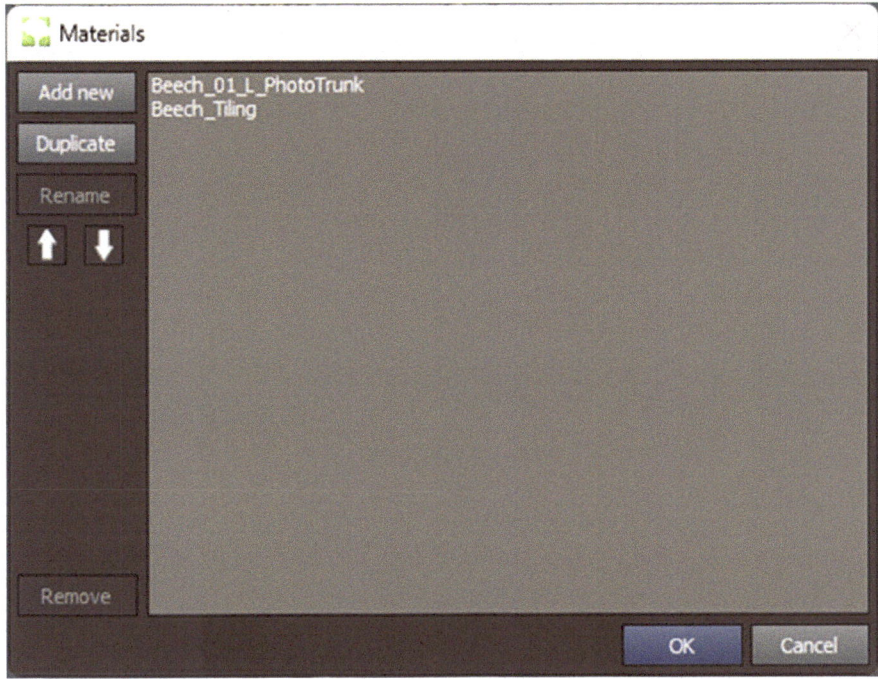

FIGURE 6.3 The materials window with two materials created.

With the photorunk material selected, you can load all the textures. If you have made your own, navigate to those. If you are using the source files, you can find the textures here: *…/CreatingHighQualityVegetation/workfiles/designer/export/T_Trunk_Beech_L_01*

Load in the Color, Normal, Height, Roughness, and Ambient Occlusion. You can do so by clicking on the corresponding texture slot, navigating where the textures are located, and selecting the file. Refer to Figure 6.4.

Then navigate to the meshes tab and select the +/– button again, select Add New and call the mesh LP_Beech_L_01, then in the High Slot load in the corresponding mesh, you can find that one here.*./CreatingHighQualityVegetation/workfiles/blender/ export/Phototrunks/Trunk_Beech_L_01.fbx*

Refer to Figure 6.5 to see what it should look like with the mesh loaded in.

SpeedTree will ask you to find associated LODs, but for the sake of simplicity, these are not included. There is a more detailed explanation about LODs in Chapter 6, If you would load in meshes in the Med and Low segments, these would correspond to LOD1 and LOD2, but for our purposes, you can select No and go back to the Materials tab.

The last thing remaining is to link the mesh we imported into the material, so with the Materials Tab Selected, select Cutouts/Meshes, select Add, and in the drop-down menu Select LP_Beech_L_01.

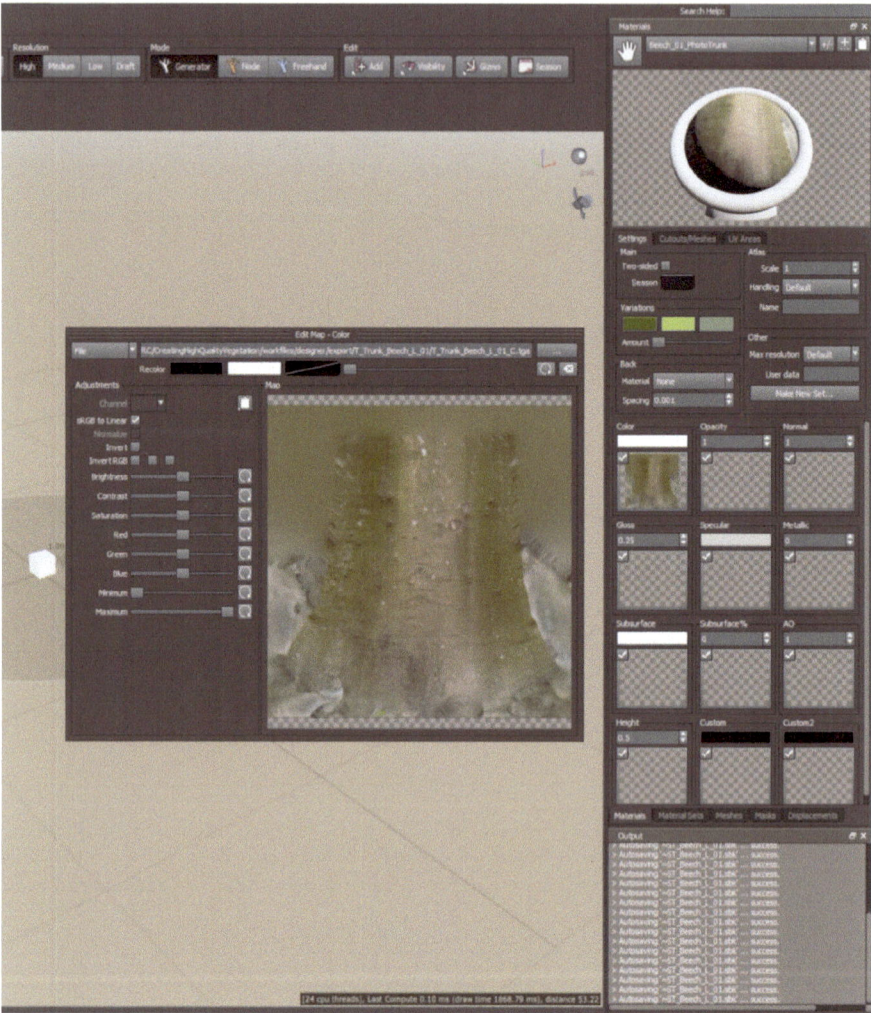

FIGURE 6.4   The texture window with the color texture applied.

Your Beech_01_Phototrunk material should now look something like Figure 6.6.

With our material set up and mesh linked, you can select the Mesh Node and browse to the Material Tab, then in the Material drop-down menu, select the material you just created; in my case, it is called Beech_01_L_Phototrunk.

It could be that you are now seeing an enormous tree trunk. The reason for that is that back in the day, SpeedTree was using Feet as a measurement, so scaling is always a bit funny even though we have imported it as centimeters from Blender.

Luckily, we can fix this inside the Mesh node, navigate to the mesh tab and turn off Use Actual Size, then play around with the Scale slider until the trunk is slightly lower than the human size scale, refer to Figure 6.7.

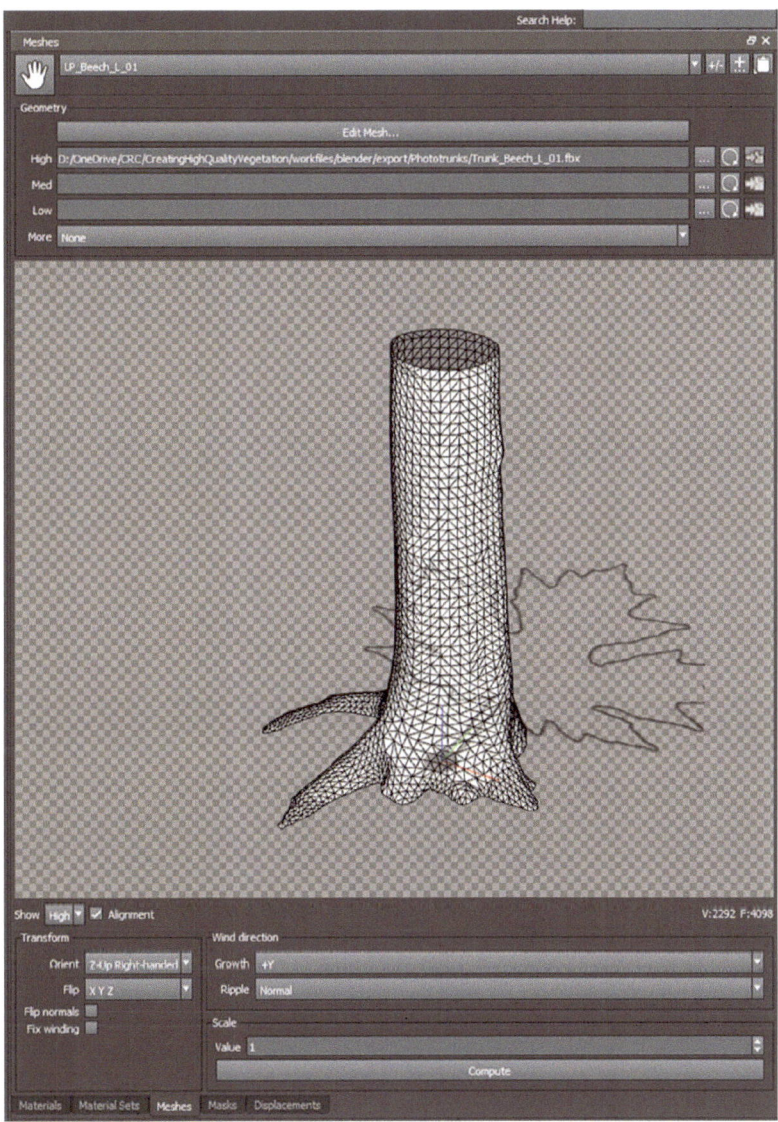

FIGURE 6.5   The photogrammetry trunk loaded into SpeedTree.

BLOCKOUT AND PLANNING

Before we get into the fun part, there is something crucial to convey; if you want this book to teach you anything, let it be this: **do not let SpeedTree decide how your tree looks**. We can either use the procedural workflow to work for us, or let it work against us.

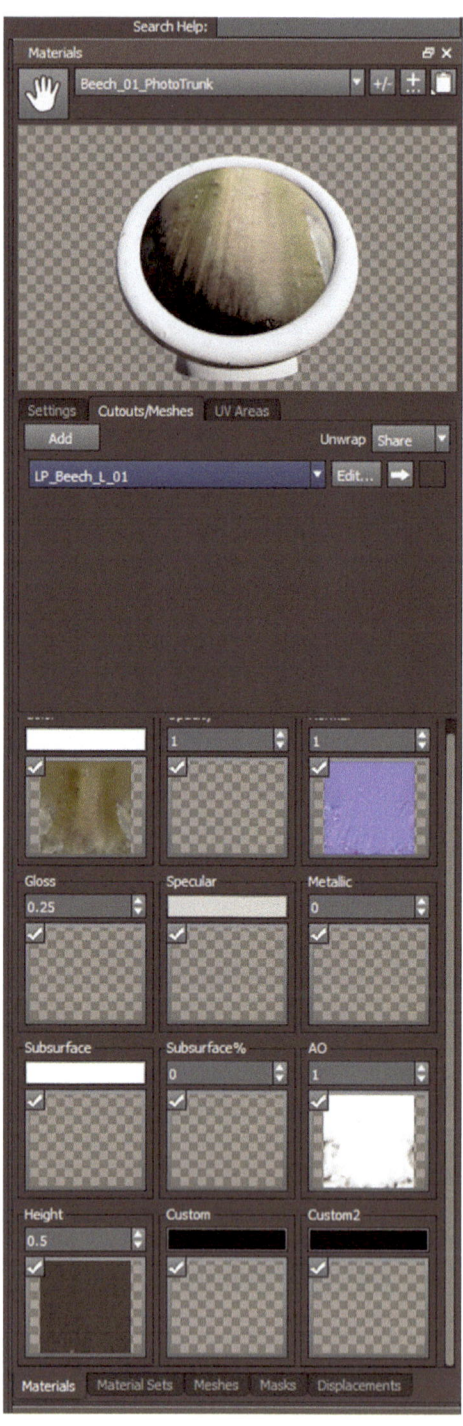

FIGURE 6.6 The material setup with a mesh linked in.

FIGURE 6.7 The speedtree UI with our trunk imported and textured, scaled down according to the size reference.

Procedural tools are very powerful, but they do come with a risk, and especially inexperienced artists suffer from this, they come with a steep learning curve, and often I see people let the tool decide the look of their tree because they fundamentally do not understand how to work with a procedural tool and fall in one of the following traps:

- Assume something looks good because SpeedTree generated it.
- Skip over fundamental steps in the process, such as a blockout and or looking at references.
- Not knowing which setting affects what and settling for the result at hand.

There are more things to consider, but these three points are, in my opinion, the most common ones. This is why I recommend that less experienced artists build manually before diving into procedural tools.

A couple of things to consider to avoid experiencing these problems is to see SpeedTree as a tool that gives suggestions. SpeedTree does not set the length of your trunk; it suggests it. It does not define the number of leaves; it gives you a suggestion; never stop asking, "Do I accept this suggestion? Does it match with what I am seeing on my reference?" and if the answer is "No," communicate that back to SpeedTree by changing the parameters.

By approaching it like that, you avoid SpeedTree making artistic choices for you.

That said, those suggestions are a very powerful tool and can help you get unstuck in a time of need; I often find myself setting up some basic parameters and then pressing the Randomize button to see if SpeedTree is giving me anything I enjoy and would like to refine further. See Figure 6.8.

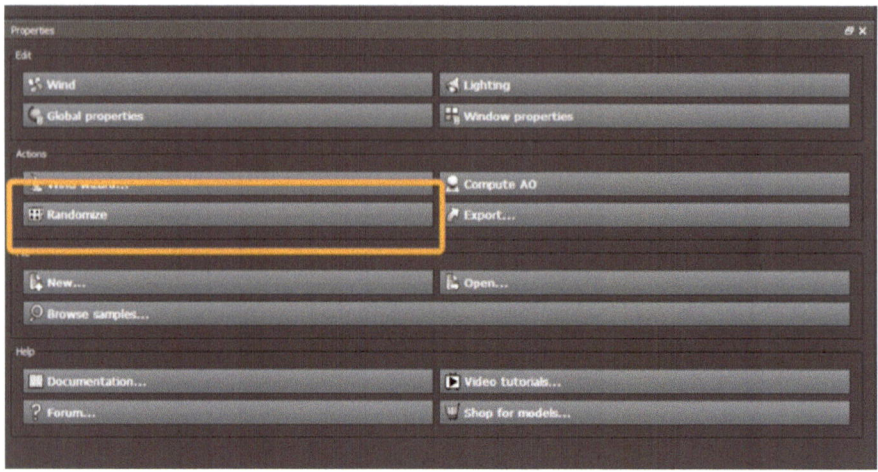

FIGURE 6.8 The SpeedTree randomize button.

Refining is the keyword, and it bridges the second point: Skipping over fundamentals.

Traditionally, a blockout refers to a block-like, simple version of what you intend to create in simpler geometry, so you can decide if you like the size and shapes before committing to spending a lot of time on the details, it's the closest thing to a sketch in 3D. But since SpeedTree generates the geometry for you, manages your scene, does your UVs, and even, to a certain degree, decides some noise and shapes as soon as you put a node down. There is a risk to diving into the details straight away; there is not a single art discipline where diving into the details straight away is considered a good thing, and making digital vegetation is no exception.

In a procedural workflow, you still have to block things out, but you approach this slightly differently, the first couple of nodes you put down are just cannon fodder, something that will be highly likely to be replaced. In traditional art, there is something called the empty-page syndrome, which is remedied by just putting down random scribbles, I often compare this step to something like that – having something to look at allows you to say, "Oh, I want the trunk a bit longer" or "On my reference, the branches seem to start higher up" none of these things are questioned when you are looking at an empty viewport wondering where to start.

This is another double-edged blade because, on the one hand, there is this common issue with creating things procedurally. Still, on the other hand, it lends itself so well for trial and error you can get things "onto paper" so quickly without making any serious time commitments, so do use that to your favor, put down a node or two, change some random settings and see what you get, if you like it, you get to keep it, and if you don't worst case you have lost a couple of minutes figuring out what you don't want.

That gets us to the third point: settling for the result at hand.

When many things are done for you, it is very easy to accept what you see, consider it "good enough," and move on. Still, I strongly recommend trusting your own voice, looking at references, and using your artistic ability to decide what you want and don't want.

There is not a single tree SpeedTree is not able to create; there could be outliers that are more difficult to do procedurally, but even for those, you can rely on the hand drawing tools and node edits – these will be used and detailed later.

As a general rule, you should gradually reduce the things SpeedTree does for you and how much you spend on details. At the start, aim for one hundred percent SpeedTree generated, using randomize to find shapes that work for you. Once you have all the elements you intend to use represented, start refining this to your liking, homing in on your references but maintaining a balance, where you have about sixty to eighty percent done procedurally, and only at the end stage allow yourself to do heavy Node edits and Freehand edits.

Lastly, before we get into making the tree, there is a final message I would like to convey.

Don't spend less time on the tree than you would manually; let SpeedTree make your work easier, not shorter. Building art takes time, and a good tree can take up to 40 hours. Doing it procedurally should not cut that time too much but instead allow you to focus on other things and push the quality bar.

SETTING UP THE TRUNK EXTENSION

We first need to extend our trunk and blend the tiling and photo trunk texture together. Speedtree has a couple of methods to do this. Under the Add Geometry, Photogrammetry drop-down menu in the Node View, you will notice three stitches. For our purposes, we want to use the "Bake Stitch," which Speedtree also recommends.

What this does is create a strip of geometry that takes part of the tiling texture and blends this with the photogrammetry trunk, this does need to be baked down so you will notice an exclamation mark next to the node, if you click on that you can see the error and sometimes SpeedTree gives you the solution, but for now, we will ignore the error in case we need to adjust settings for the stitch later. We will bake it at the end when all our edits are finalized to avoid having to rebake the stitch multiple times.

When making the Stitch, SpeedTree will automatically make a Branch node as well, refer to Figure 6.9.

Additionally, you will see a bunch of other tabs, we will go over these in more detail later in the chapter, but it could be worth to familiarize yourself with them a little before that, to summarize the ones we will be looking at in this chapter:

All

This tab will combine all of the settings into a long list, I find this useful when I know the name of the setting I am after, but unsure in which tab it is. In these cases, I switch to the All tab and use the search bar at the top.

FIGURE 6.9   SpeedTree with the stitch node and branch node.

Gen

It contains all settings regarding generation, the number of branches, where they start and end, which rotation, position, and scale they have, and how they relate to the parent branch.

Spine

This tab controls the curve, spline, or spine of the branches and is useful for changing the length and orientation, as well as some physical attributes, such as pruning or adding potential noise to the branches.

Skin

All the settings for the geometry can be found here. The thickness or radius can be changed, as can branch welds and splits – basically, anything related to the visual part of the tree.

UV

Self-explanatory, this is where you make changes to your UVs.

Displacement

We will be using some textures to displace our branches, any settings as to how this texture behaves can be found here.

Material

We will set up a couple of materials, and they are applied to the tree here,

Now, let's get back to the stitch.

Your stitch could be a bit off-center. To fix this, select the stitch node and switch to Node in the top right, allowing you to make hand edits. The stitch is an exception to hand edits as it relies on user input to define the center of the trunk. Use shortcut W to bring up the move gizmo and center it above the hole so the extended branch aligns nicely with the photogrammetry part.

I called the Created Branch node "Trunk," and in the Material Tab for this node, we can apply our tiling texture; we have previously created the material entry for the tiling texture, so all you have to do now is link in the corresponding textures, this process is identical to the photogrammetry trunk material we set up earlier in this chapter. You can find the texture I used here: *../CreatingHighQualityVegetation/workfiles/designer/ export/T_Tiling_Beech_L_01/*

In the Spine settings of the Trunk Node, you can now roughly define the length of the tree; in my case, I have made the trunk about 42 units high. You can see the units in the viewport at the top of the ruler.

The thing with the default nodes is that they create perfect tube-like structures; this is a giveaway that the tree is built in SpeedTree and looks unnatural. Most trees will have some deformations, usually contextualized, but that is harder to achieve procedurally, so we will have to strike a balance between general imperfections and ones that come from an event in the growth of the tree (new branch growth is an excellent opportunity to do this).

Have a look at Figure 6.10.

You will notice in this reference that whenever a new branch grows out, two things happen: first, it creates a bit of extrusion, and second, it makes the "eyes" we observe in beech trees. Other trees do this as well, and an eye is usually a scar from a fallen branch.

But marked in orange, you will notice this is not a hard rule. The trunk deforms here and there without an immediate explanation as to why. That is the type of deformation we are after, which can be achieved using Displacement.

CREATING A BRANCH STRUCTURE

Navigate to the Displacement tab and play around with these settings for a while, trying to strike a balance between being visible enough from a distance and not so aggressive that it looks unnatural. I like to think of these things as things you feel rather than see.

FIGURE 6.10 Reference of a beech tree in winter.

In my case, I set the displacement power to 0.28 and played around with the Shape parameters like Amount, Twist, and Scale to create a subtle twisting deformation across the trunk.

You can refer to Figure 6.11 for how much it should be displaced.

With that setup, we can go ahead and make a new branch node, Right-Click in the node view, Add Geometry, and select branch; you should end up with something simple, as seen in Figure 6.12, something very powerful in SpeedTree is the ability to copy settings from one node to another, negating the need to set things up repeatedly.

You can try this with the displacement settings set in the Trunk Node. Navigate back to the trunk node's displacement settings, Right-Click the Displacement tab, and say Copy, then paste this into the newly created branch node to have a similar displacement on the branches as we have on the trunk. Don't be afraid to edit this a little, as smaller branches usually deform more aggressively than the trunk. Refer to Figure 6.12.

If you haven't already, now is a good time to bring in your reference so we can start shaping the tree accordingly. You can find the reference board I used here ..\ *CreatingHighQualityVegetation\ref\RefBoardCreatingVegetation.pur*

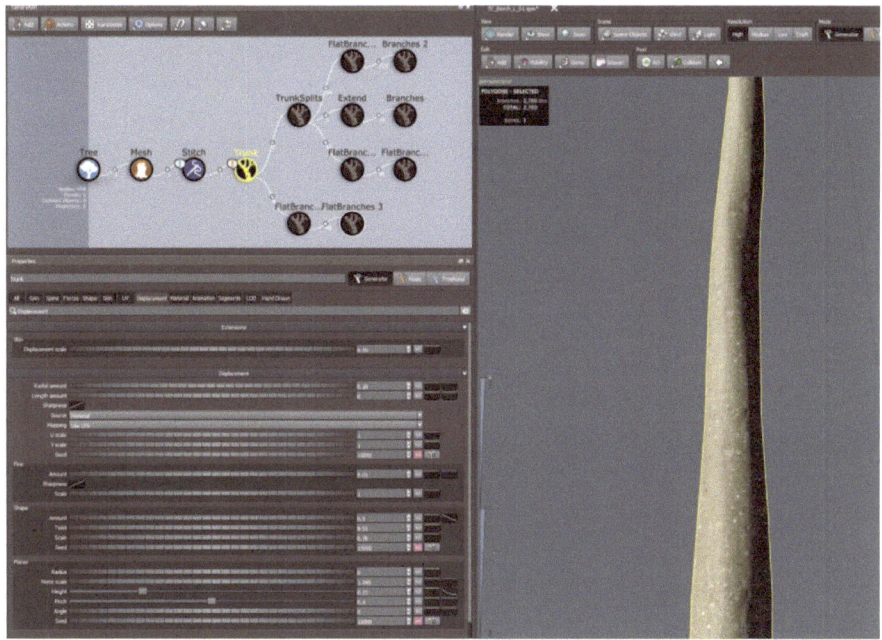

FIGURE 6.11 SpeedTree with an example of a displaced trunk.

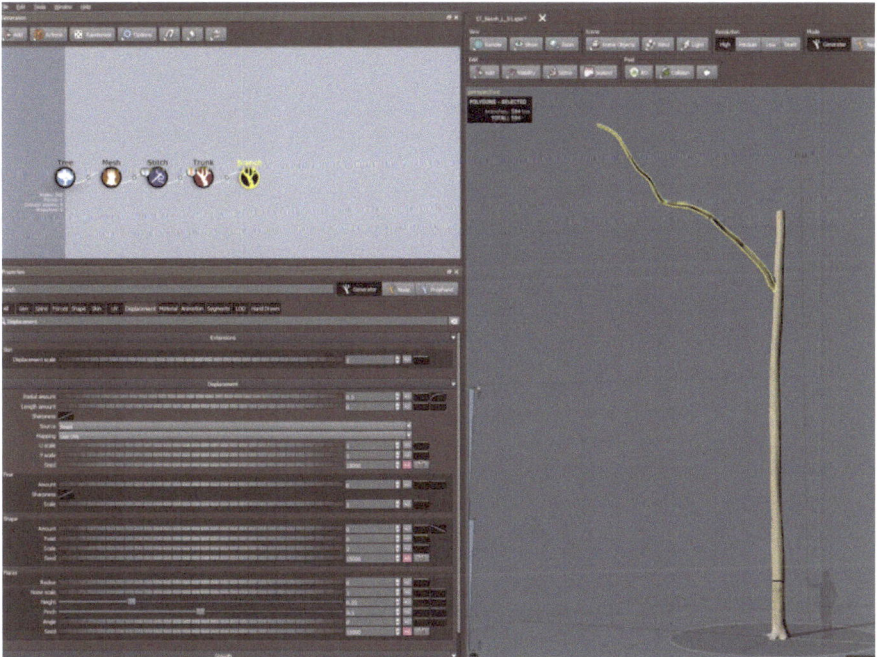

FIGURE 6.12 SpeedTree with an example of a branch added.

You will notice this is split up into general references and Main References. Although it is good to have many references to look at for inspiration or perhaps grab a unique feature or two, in these early stages, I like to look at one or two images at most to define all my major shapes.

Generally, I like to build the largest shapes first; in this case, those are the trunk splits, and I have named my Branch Node accordingly. Refer to Figure 6.13.

FIGURE 6.13 SpeedTree (left) with our branches shaped according to the reference.

There are a couple of things worth noting: the phyllotaxy is hard to see, but in the Gen tab, I have set Mode to Phyllotaxy and the Style to Alternating. Additionally, look at the number of branches splitting off and get into a similar ballpark using the inter-node length in the Gen tab.

Continue to look at how the branches come off the trunk. They almost immediately go up; this can be achieved using the Start Angle in the Spine tab within SpeedTree. I have set mine to 0.143.

It is essential to point out that you should not aim for perfection at this stage. Just focus on getting all the elements in there. Should the branches be thicker? Or longer? Maybe, but the sum of all parts will decide the final look, so we are just trying to get it into the right ballpark – we can tweak it all we want later.

Additionally, we can observe that there is some sort of noise happening primarily toward the end of the branches. Luckily, Speedtree has that function precisely and can be found in the Spine tab. Under Noise, change the amount of Late noise. I set mine to 0.7.

Look at your reference to see if that value makes sense to you.

With these larger shapes, where we are only dealing with a couple of branches, it is much more manageable and possible to edit individual branches, so this is a good opportunity to get familiar with the Node mode in SpeedTree.

You can switch modes at the top of the viewport; there are three modes: Generator, Node, and Freehand. Up until this point, we have been using the completely procedural

Generator mode, but if you switch it to Node, you can select individual elements in your tree and adjust them individually. Refer to Figure 6.14 for reference.

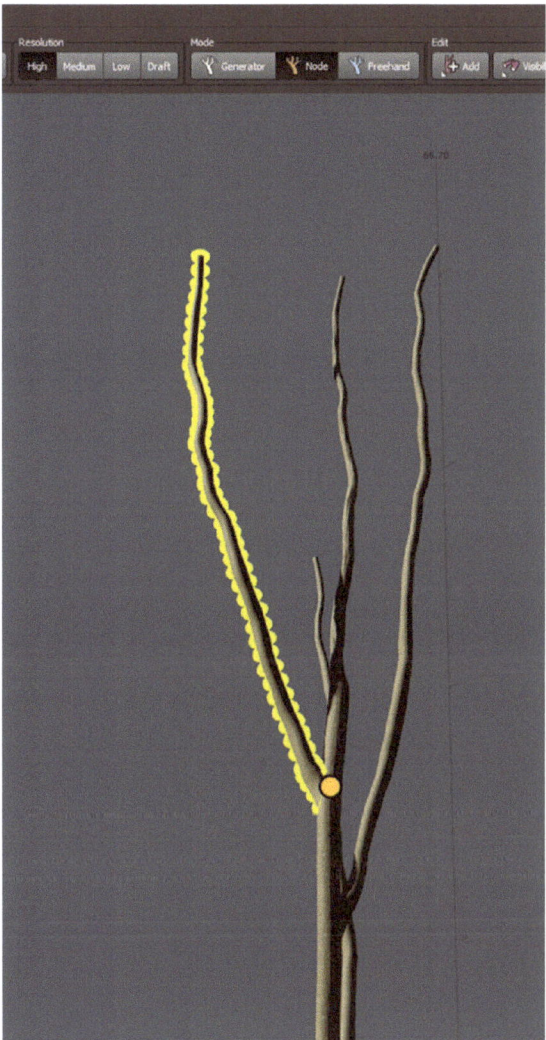

FIGURE 6.14 The node edit mode with a branch selected.

This does come with the disclaimer and reminder of what I explained earlier about not using Node editing in a blockout stage: using the Node mode on a generator that has about four branches is not something you would not be able to easily redo in case you make a major change to the node, so even though the tree is still very early on. You want to rely as much as possible on procedural generation; the bang for the buck of doing it in this step is high and, therefore, acceptable. As you grow as an artist, you will find that it becomes easier to decide if it is okay to do some hand edits, but for now, just consider, if you would have to throw it all away, would it be a huge deal or just a couple minutes of rework? If you feel like it would be a huge deal, you have probably relied too much on hand edits.

Play around with this a bit and get comfortable with how it behaves. Using the Node Edit mode, get your branches to look similar to your reference or Figure 6.14, make the finishing touches using the Node mode, and then switch back to Generator mode.

It's not a tree without some more secondary branches, so we can start looking at that. These will, to a certain degree, define the shape of the canopy, so we have to consider some things.

1. This is a tree that grows in the forest, meaning it competes for light with other trees. Its common behavior is to grow up, meaning that the apical part of the canopy is likely pointing upwards. This is supported by reference.
2. The reference also shows that many of the Beech branches tend to be almost horizontal, creating platform-like structures.

Then, in between, there are somewhat diagonal branches, with the tendency to grow up the higher up they are in the tree.

When building a canopy, I liked to divide it into three segments: the top part, which I refer to as the "crown," the sides, or "body," and lastly, the bottom part, which I refer to as the "skirt."

I find it easiest to start with the crowns, as they are somewhat of an extension of what we already did. Referring to Figure 6.15, you can see what we are after.

FIGURE 6.15   The crown of the Beechtree is visible in Speedtree and reference.

To achieve this, I added another branch generator by right-clicking on the generator view and selecting Add Geometry, Branches. I have renamed these "Extend" as I use them to extend the trunk splits. To do this, go to the Gen tab and set the number to 0 – this will temporarily remove all branches. In order to get them back to just the extends, set the Extend Parent Type to "Any" also in the Gen tab.

You could also use this node to affect the whole canopy, but I find dividing this up into a crown-specific structure within Speedtree gives me more control when tweaking it later. The trunk branches will also be the root for different parts of the canopy, and if I want to make changes to specific areas, I like to keep those separated in my node graph.

With the extensions built, we can add another branch generator and keep this relatively simple, I set mine to a Number of 9 in the Gen tab and made sure they are pointing upwards using the Start Angle; additionally, I added some noise toward the end, like we have done in previous steps and for now that is enough to move on to the next step of the canopy, the body.

The body is usually one of the largest parts of the canopy, but with beeches, I feel like the body and skirt are equally important. The body serves as a gradient from the upwardly orientated branches at the top to the almost horizontal ones toward the bottom.

Again, while working on this, continuously look at references, see Figure 6.16 to see what we intend to build and what sort of reference to look at.

FIGURE 6.16   The body of the Beechtree canopy is visible in Speedtree and reference.

With the majority of the body out of the way, you can move on to the next step; currently, most of our branches are pointing up, resulting in a very uniform-looking tree; in nature, this is hardly ever the case and beech trees have in many cases very horizontal looking branches, this is visible in most reference and need to be recreated. Refer to Figure 6.17.

FIGURE 6.17   Horizontal branches visible in Speedtree and on the reference.

We can initially copy over the previous set of branches to build this. Please work with the start angle to get them as flat as possible and potentially reduce or increase the gravity to ensure they are not pulled down or pushed up.

On top of this, there is a very powerful feature in SpeedTree called Forces – Forces are parameters from outside that affect your tree. To add forces, navigate to the Scene Objects Tab, Forces, and then Add Force, refer to Figure 6.18.

You will find a handful of forces here that are mostly self-explanatory, and I do encourage you to play around with this a bit if you have never worked with them before; the one we are interested in is the Planar Force, so go ahead and add one, with the newly created branch nodes selected. This will add a new object to the scene that can be selected; I have set mine to a Strength of 4.3 – essentially what a planar force does: it tries to flatten branches, which is perfect for the result we are after. Refer to Figure 6.19.

When you apply a force, it could be that it has applied to all your branches. If that is the case, select a node and navigate to the forces tab. In the forces tab of each node, you can choose which forces are applied and which should not be applied so that you can get very granular in your approach here. For now, we only want to use our Planar Force for the branches we just created, so make sure to uncheck it everywhere else.

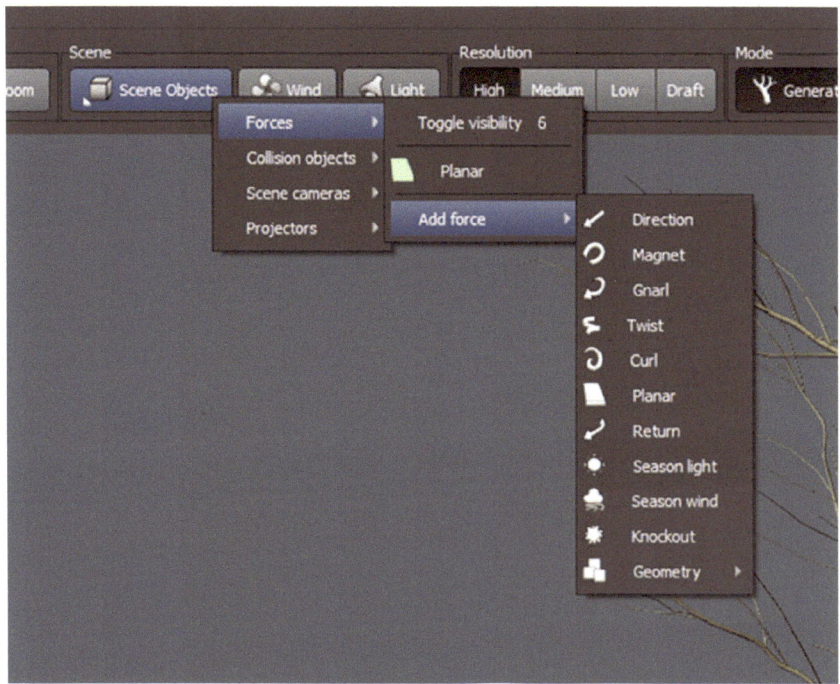

FIGURE 6.18   The add force tab in SpeedTree.

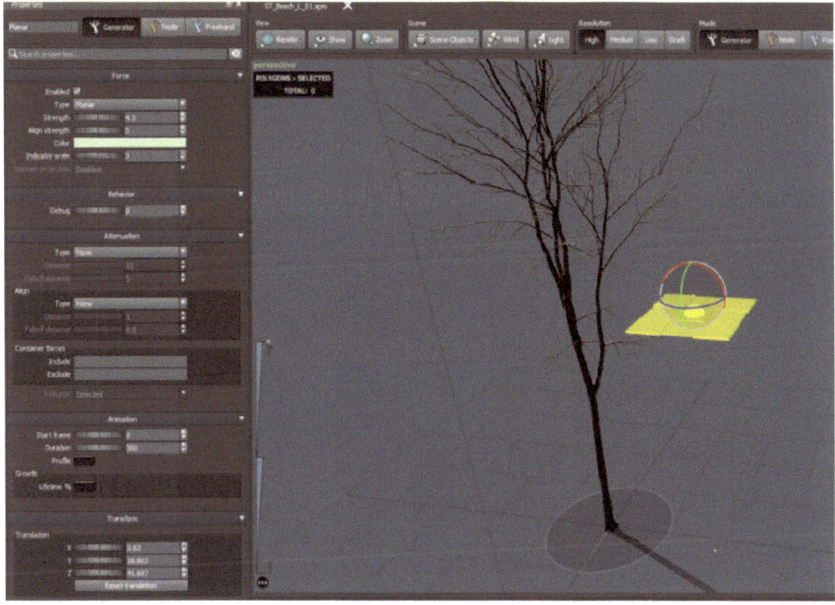

FIGURE 6.19   The planar force in the viewport in SpeedTree.

For the time being, this gives us all the elements we need in the upper canopy; feel free to play around with the elements just created and see if you can bring it closer to the reference, it is hard to convey in a book how much iteration is required but I am constantly rotating around my tree and comparing it to the reference to make sure I am honing in on what I see on the reference.

It is essential to rotate around the tree as it needs to work from all angles. On that subject, it is a good idea to see if you can get some shape variety on each 90° increment, so if you look at the tree from the front, it looks like a different tree than from the side. That way, if you repeat them in the engine later, they will bring more variety and look less repetitive.

Notice in the reference that there are also thinner branches, shooting off directly from the trunk, that look like the branches we just created. Refer to Figure 6.20 to see what is meant.

FIGURE 6.20   The flatter branches coming off the trunk are highlighted on the reference and visible in SpeedTree.

We can once again reuse some of the structures we used before for the flatter branches, but this time, copy them and attach them to the trunk directly. Use the First, Last, and Position values in the Gen tab to ensure they spawn only toward the bottom of the trunk and adjust the planar force in the force tab accordingly. Again, compare these to your references to get an idea of how these should look.

These branches are very thin in the reference, so you should probably scale them down more than you think. Keep an eye on this throughout the process; this is something to look out for as well when importing the tree in Unreal Engine later.

With all this in place, play around with different factors a little. We now have all the major elements seen in the reference setup, and this is a good opportunity to do some iteration. Save a copy of your file in case you want to go back, compare it to your references, and make incremental changes,

Make sure to familiarize yourself with the Art Director Gizmo that appears when you click on a branch in the viewport. It can change the Start Angle, Gravity, and Noises very quickly, and knowing how these work can save you a lot of time. Refer to Figure 6.21 to see the gizmo.

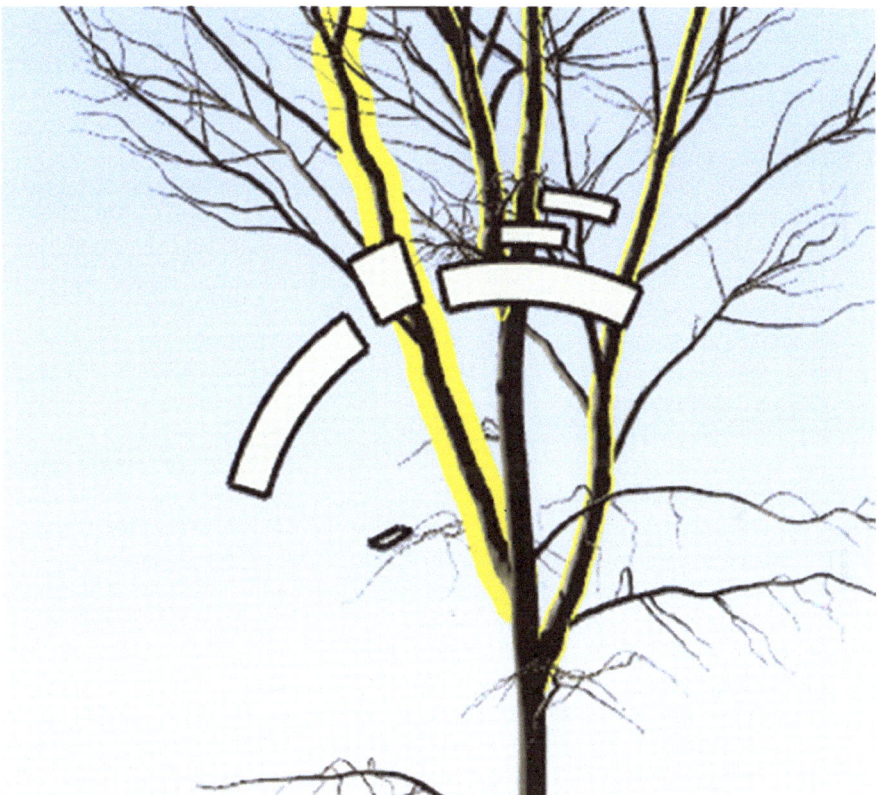

FIGURE 6.21   The gizmo to edit branches quickly within the viewport.

Try and press randomize a couple of times, this usually gives you a quick reality check as it allows you to look at different iterations of the rules you setup, for example in my case, I noticed the trunk splits where starting a bit too early on the trunk, and I changed the position again to push them further up.

Once you are happy with your result, we can move on to building the Leaf Clusters.

BUILDING THE LEAF CLUSTERS

What are leaf clusters? Leaf clusters are one of the many terms used for what we will create, and they are primarily used in SpeedTree. Outside of SpeedTree, they can often be called leaf cards, canopy cards, or branch cards. The reason is that they are essentially a branch and leaf texture baked down onto a plane to save on geometry, and they end up looking like cards.

Making a good cluster or card is essential for the final look of the canopy. Carefully crafting the texture can avoid or remedy many problems, so I recommend spending as much time as possible to get something that looks nice and volumetric while not giving away that it is made of simpler geometry.

Exciting developments are happening on this subject at the time of writing the book, and we are already partially moving away from the card workflow. In the near future, this workflow will most likely become completely redundant. And be replaced by an entirely mesh-based approach, so instead of baking down clusters, we will have individual leaf meshes all over the tree and most likely no longer rely on opacity maps but cut out the mesh entirely. The techniques shown in this book are used for both methods and should remain relevant for the foreseeable future.

We will be creating our card clusters in a separate SpeedTree file, so go ahead and click File, New, and select a blank file. For this file, I also went to Window Properties and changed the Background Style to Solid Color, as I think the gradient makes it harder to judge shapes and sizes.

Start by putting down a trunk node by right-clicking in the graph view, selecting new geometry, and selecting trunk. Then, do the same for a branch node and make sure it only has one branch.

For this branch, go to the Spine tab, and under Noise, set everything to 0; in this step, any deformation is unwanted. You can do the same for the trunk; it is a bit neater. We will not be using the trunk in the final result, so it is best to have it have as little effect as possible. Then, lastly, use the Start Angle in the Spine tab to make the branch come out as straight as possible; for me, a value of 0.663 did the trick.

For the trunk, under the Displacement tab, set the flares to 0, then navigate to the Skin tab. Under Radius, you will see that the Absolute value is set to 1 by default, and it has two curves. Whenever SpeedTree shows you two curves, the first one is related to the length of the object you have selected, and the second one is relative to its parents; in this case, there is no parent, so we can ignore that, adjust the curve so the thickness is the same across the whole trunk. Scale it down; I have set mine to 0.25. This turns the trunk into a basic-looking cylinder.

You can refer to Figure 6.22 to see what this is supposed to look like.

On a side note, curves are incredibly powerful, and it is a good idea to play around with them; you can pull the curves around very extremely and get a lot of control over the shaping of the branch, at this point, it is a good idea to take a couple of minutes to play around with the curves to get a good understanding of how they work.

Since we will be baking this down, it is best to look at it from the angle from which it will be baked. In our case, that is the *XY* plane. In the top left of the viewport, you will

FIGURE 6.22 The trunk and straightened-out branch created for the cluster workflow.

see Perspective; you can click this and select the XY plane to look at what we built from above. Visible in Figure 6.23.

Alternatively, a better approach is to create a new camera by going to Scene Objects, Scene Cameras, Add Camera, and select Orthographic and place this above your intended leaf cluster, this does increase the scene complexity so I recommend only doing this if you are slightly familiar with SpeedTree. Feel free to play around with this setup and see which route you like best.

For the branch, you can set the rotation value in the Gen tab to 0.25 (90°) to ensure the branch is pointing up. Select the trunk and press H to hide it; make sure only the branch is visible so the trunk will not appear in any of the textures later. To ensure we know what will be visible in the bake, click anywhere in the viewport to see the global properties window and select Window Properties. This is also where you change the background color, but this time, turn on Show under Screenshot Safe Frame; anything within these borders will be baked, so we need to stay inside these lines to ensure a smooth bake.

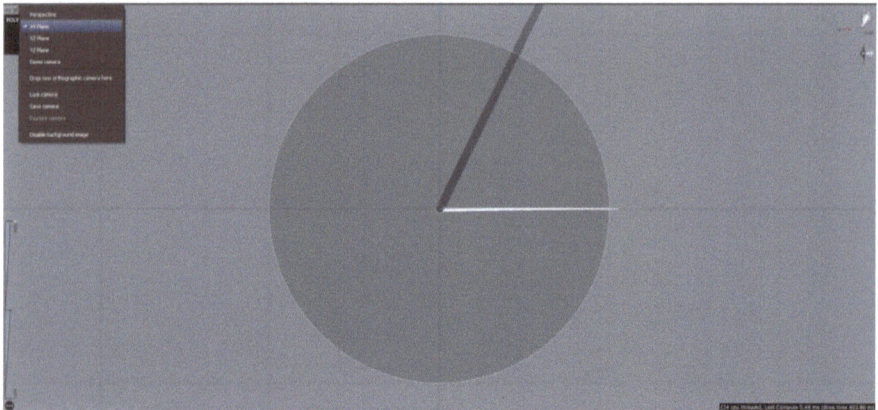

FIGURE 6.23   Viewport seen from the *XY* plane.

My branch ended up being a bit on the long side, so I changed the length in the Spine tab. I set the % of parent to 0 so it no longer looks at the trunk and adjusted the Absolute value.

Now is an excellent time to bring up the PureRef board with references and look at the Canopy Card References. We will need to build a couple of branches, but I like to start with the largest one to derive smaller branches from that.

The PureRef board details the reference. Refer to Figure 6.24 to see what references I am examining and discussing in the book.

Please take note of a couple of things. Most notably, the leaves spawn in an alternating phyllotaxy, which tells us the Generation mode should be set to Phyllotaxy for the leaves later. They spawn at the end of the first and second-generation branches and in clusters on the offshoots. When they spawn in clusters, they usually spawn in groups of 3. These pictures were taken in early spring, so most leaves are similar in size, but we see a slight scale difference toward the tips.

These observations give us an outline of the ruleset we would like to build.

Let's start by changing the radius to be thinner and switching to Freehand mode to shape the branch precisely as we want it. It is less troublesome to use Freehand Mode for creating clusters, as you usually want to have very specific results when building clusters. They are achievable procedurally but will be much harder to control and maintain.

If you switch to Freehand mode, a new gizmo will appear. I recommend playing around with this gizmo until you understand how it works, if you hover your mouse over an element in the gizmo, it will explain what it does. Refer to Figure 6.25.

Comparing the shape to the reference as you go along, in this step, it is also important to remember that we are not trying to copy everything one to one, I recommend just doing the rough outlines, as we would like to get something on our tree as quickly as possible, then refine from there. At this stage, we are still unsure if the shapes we build for the clusters will look good when placed on top of the tree, so it is best not to commit to anything just yet. Refer to Figure 6.26 for a reference of our primary branch.

Right-Click in the viewport and select Add Geometry, Branch, and Big Branches, by default These will spawn around your tree; you can increase the amount to have

FIGURE 6.24 References in the PureRef board.

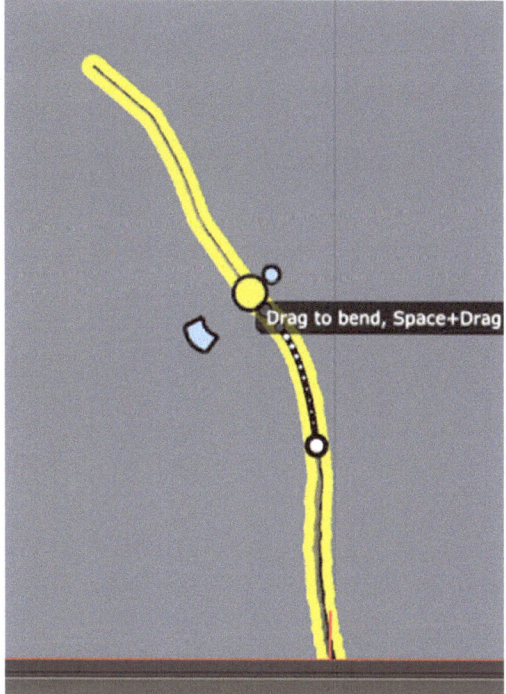

FIGURE 6.25 The freehand gizmo with the tooltip exposed.

FIGURE 6.26 The branch we are recreating on the left (highlighted in blue), and the SpeedTree version on the right freehand gizmo with the tooltip exposed.

enough branches to work with. Then change the Phyllotaxy to Alternate, this should flatten them out a little already, but you can use the Node Edit mode to select each branch and the Rotation value to flatten these out more and, move them in place, I have not used the Freehand mode for the location of the branches, just adjusted the rotation, as I try to keep things as procedural as possible while maintaining precise control over where things go. I would also like to point out that when working in SpeedTree, there are always multiple solutions to the same problem, so if you are feeling adventurous, I recommend trying out different things to get these results, perhaps a flattened force similar to what we used on our tree branches will work wonders here, or another phyllotaxy setting will give you better or similar results.

Move on once your result looks similar to Figure 6.27.

Now, we have our first- and second-generation branch setups. These largely decide the shape of our branch cluster. Moving forward, I will use less of the Freehand and Node modes and rely more on a completely procedural setup. This also means it will be harder to follow the references one by one. We will rely more on the rule set defined earlier in this chapter.

Right-Click in the node graph, but this time select Add Geometry, Branch, Little branches. This should give you a bunch of little branches spawning on top of your second-generation branch. Go to the Gen Tab, set the Mode to Phyllotaxy, then set the Arrangement Style to "Alternation(distichous)." This will make sure they spawn flat.

If you look at the references for these branches, you will notice that not all of these third-generation branches are similar in length, but an obvious rule does not seem to be attached to them. This has multiple reasons. Firstly, a tree will always try to grow

FIGURE 6.27 Secondary branches highlighted in blue on the left, and the SpeedTree version on the right.

as efficiently as possible, so if a branch is not needed, it will not grow, and vice versa. Secondly, in nature, you are also dealing with branch breaks due to various factors.

We can simulate this effect in SpeedTree using multiple settings. The most obvious one is length; if you go to the Spine Tab and have a look at the % of parent setting, you will notice it has three parameters attached to it: the default value, a +/− sign, and the graph, we are interested in the +/− sign, this allows you to add some variety to the initial value, I set mine to 0.1 meaning it can be ten percent larger or smaller than the value I put in the first slot. Refer to Figure 6.28.

Additionally, we notice that they come out of the second-generation branch at quite a harsh angle. We can change the Start Angle in the Spine Tab to achieve this. Alternatively, you can go to the All Tab and search for these settings. This search bar becomes increasingly useful as you get more familiar with SpeedTree.

All that is left now is to deal with the branch breaks. The benefit of using branch breaks is that when we spawn leaves on this later, the leaf generator will ignore the branch breaks, an effect we are after to make the branch look more natural and realistic.

You can find these settings in the Spine Tab under the Prune segment. Normally, these settings are used for a tree that stands up right, but in our case, we are working on the *XY* plane, so you will find not all the Prune settings work as intended. I found using just the Interior Prune settings works best, and I used a value of 0.087.

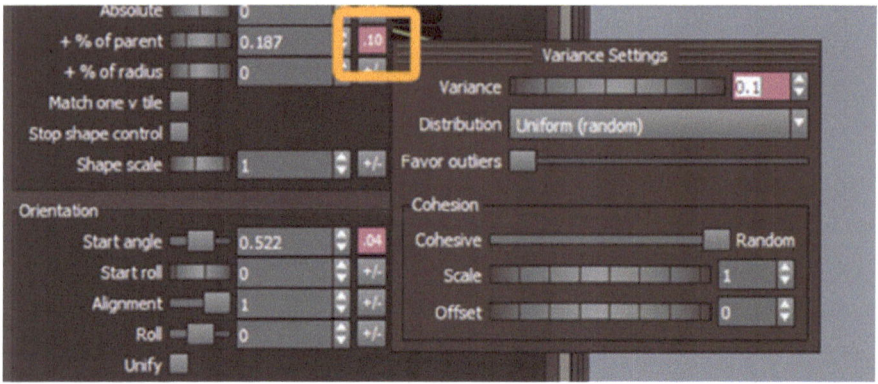

FIGURE 6.28   The variance setting highlighted in orange.

Additionally, in the Spine Tab under break Settings, we can increase the chance of a branch breaking; in my case, a value of 0.65 was used, adding some additional break up to the structure and taking it a bit closer to the reference. You can refer to Figure 6.29 to see the result of these settings compared to the reference.

FIGURE 6.29   Left shows the reference, and the right shows the speed tree viewport with tertiary branches highlighted.

Worth noting is that when building branch clusters with the intent to bake them down into a texture, it does not matter how many triangles we use, the reason for this is because we will combine all of these triangles into a texture that gets applied to a new mesh with much less triangles, so when building the high poly branches, feel free to ignore the total number of triangles.

I recommend checking your Segment Tab and adding enough segments to all your branches to allow for much detail to be packed into the mesh if you use too little the branch will end up looking cheap, if you select the Tree node in the Graph Editor you will be able to see the total tri-count in the top left, I am using around 23.000 triangles (tris) for just the branches, but depending on the complexity this can quickly go up to 100–200 thousand, do not shy away from adding more geometry, in general, I like to focus on quality first and performance after.

Lastly, we can copy our previously created Little Branch Node and add it to itself, creating some branch splits and offshoots, generally making the branch look more natural. Again, always look for ways to add some additional details you see on your reference; the more, the better. You might have to edit the amount a little if you end up having too many branches spawning, but other than that, the settings remain the same as the previous node.

That said, this node concludes our branch structure, so feel free to go back and forth between the nodes and make any changes you deem necessary; double-check with your reference and see if anything is missing or does not quite look similar. Do not let perfect stand in the way of good—as always, we are not trying to get everything perfect at the moment but get a good number of elements in so we can tweak the setup later. It's all about iteration.

Lastly, copying materials from one SpeedTree file to another is possible. In the ST_ Beech_L_01 file, navigate to the Materials tab, use the Copy button at the top right to Copy the Beech_Tiling material, and paste this into the Material Tab of the Cluster file.

Add the Material to all the branch meshes, and like how the Displacement for the tree was set earlier in this chapter, set a slight displacement for the branch clusters.

Refer to Figure 6.30 for the location of the Copy button.

While in the Material Tab, create a new material called Beech_Leaves_S and use the textures found here: *../CreatingHighQualityVegetation/workfiles/designer/ export/T_Leaves_Beech/*

To populate the corresponding slots, these leaves had a similar treatment to generate the Normal, Roughness, and subsurface map as the textures created in Chapter 6; if you want, you can also use an online library or create your own leaf textures.

In this material, in the settings tab, make sure to set it to Double-Sided, as we want the leaves to be visible from both sides. If you do not set it to Double-Sided, you will look through the mesh from one side, and the texture will only be visible from one angle.

An important thing to notice is that SpeedTree works with a Gloss texture instead of a Roughness texture; these maps do the same thing but with inverted values, so double-click the Gloss Texture and select Invert to make sure it behaves correctly in SpeedTree.

The meshes for the leaves can be entirely created within SpeedTree using the Cutout editor; in the Beech_Leaves_S material, go to the Cutouts/Meshes tab, then use the Add

FIGURE 6.30   The Copy button in the trop right corner of the Materials Tab.

button to create four mesh slots, then navigate to the Meshes Tab and use the ± button to manage the meshes. In that window create four new meshes called Leaf_S_1, 2, 3, and 4. With the meshes created, navigate back to the Materials Tab and to the Cutouts/Meshes tab.

The setup should look like Figure 6.31.

If you press the Edit button next to the mesh inputs, it will open up the cutout editor for that specific mesh; by default, it should look similar to Figure 6.32.

There are many things you can do in this editor, but for the time being, we are only interested in getting the points in the right location and setting the angle of the leaf.

If you refer to the numbers in Figure 6.32, you will see the following:

1. The point settings, there are four of them: from left to right, the Move Points single, Move Points multiple, Remove points, and Reset all Points; I usually press Reset Points first to start with a clean slate; points, in this case refer to the vertices of the geometry created.
2. The points in the viewport, the points will automatically create geometry in between them.

FIGURE 6.31   The cutout and meshes tab with four meshes setup.

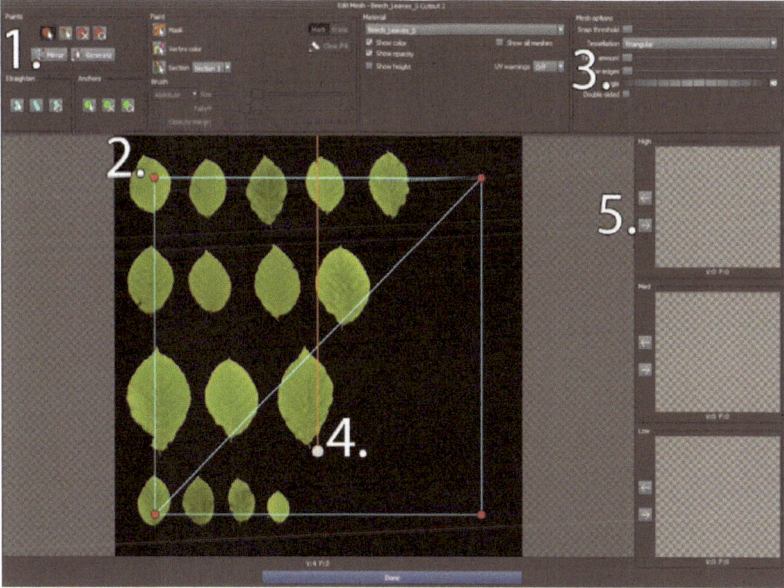

FIGURE 6.32   The cutout editor in in it's default state.

3. The angle slider. With this slider, you can change the leaf's orientation.
4. The angle gizmo, the orange line should follow the overall direction of the leaf.
5. The mesh slots: Once you are done placing points, you can select the Right arrow to load them into the corresponding slots. High, Med, and Low refer to LOD0, 1, and 2.

You can use the mouse scroll wheel to zoom in and out of the viewport, zoom in on the small leaves at the bottom, click Reset All Points, and move the angle gizmo to where the pivot is supposed to be, then use the Angle Slider to align it with the leaf center vein as good as possible.

Refer to Figure 6.33.

1. The angle gizmo moved and rotated to align with the leaf as much as possible.
2. Using the Move/Point tool to click around the leaf.
3. Use the Move/Place Point tool to create points on the center vein. This ensures that when we fold the leaf later, it respects the center vein.

FIGURE 6.33   Steps taken to create a leaf mesh.

4. I usually add a small amount of tessellation using the Tess. Amount in the top right of the cutout editor: this creates extra geometry in between the segments, which will help ensure any deformations to the leaves will look smooth and not show any geometry; again, since we are baking this down later, it does not matter how much geometry is added.

Once you have your points placed, make sure to load this in all three mesh slots. We will not be using the Med and Low slots, so you can skip these. That said, I always like to add them, just in case. You can use the same geometry for all three slots. Refer to Figure 6.34 to verify that your editor looks correct.

FIGURE 6.34 The mesh slots filled with geometry created using the cutout editor.

Spend time familiarizing yourself with the editor and repeat these steps for all four small leaves at the bottom of the texture sheet.

If you are done with the Small leaves, go to the Materials Manager and duplicate the material two times, rename these to Beech_Leaves_M and L for Medium and Large, set meshes for these similar to how they are set for Beech_Leaves_S and use the cutout editor to create geometry for the Medium and Small leaves as well, I have done them all, but about three of each should give you enough variety.

We need three materials to tell SpeedTree to spawn smaller leaves toward the end of the branches or in other specific places; if all the meshes are part of the same material, it is harder to control which leaves go were.

With our leaf meshes created, we can spawn them on the branch structure and finalize the large branch for our texture atlas.

To do this, Right-Click in the graph node and select Leaf. You might notice there is also a Batched Leaf. We are not interested in that one. A batched leaf is used when spawning a large number of leaves, but if we use that one, we are no longer allowed to make edits to individual leaves. It is, however, a lot quicker to load for SpeedTree, so it is incredibly useful if you are building a tree with individual leaves instead of a baked-down cluster.

Add the Leaf Node to the Little Branch node first, and in the Gen tab, set the Mode to Phyllotaxy and the Arrangement Style to Alternating (distichous), in the Skin Tab, scale them down to a size that makes sense according to your reference, in my case I set the Size to 0.261, then go back to the Gen Tab and change the Internode Length to make sure they spawn in clusters of 3–4. In my case, a value of 0.02 worked well. Lastly, in the Orientation tab, I have set the Sky influence and Align value to 1 to ensure they align with the branch properly.

Sky Influence controls how much the leaves will align toward the sky; higher numbers will result in the Z-axis of the meshes pointing up toward the sky. Additionally, the Align value rotates the tip of each leaf or leaf cluster toward or away from the parent, setting it to 1 will result in a look of continuous growth.

You can also add some variety in this using the Variance Settings. Double-check with your reference to see if this makes sense.

At this point, you might notice that you have a lot of overlapping leaves, similar to Figure 6.35.

FIGURE 6.35   The SpeedTree viewport showing the current leaf setup.

There are a couple of ways to fix it. One way is to increase the internode length or use knockout to remove some leaves, but with the Phyllotaxy mode, you have little control over the absolute amount of leaves, so I find that this way, you have to increase the internode length too much and it would move away from how it looks in the reference; additionally you can switch to Node mode and remove and rotate the leaves that are causing issues. But, there is a feature in SpeedTree that helps a lot with this out-of-the-box, Collision.

The collision settings are in the top right corner and turning them on will help a lot already. Refer to Figure 6.36 to see where the settings is located.

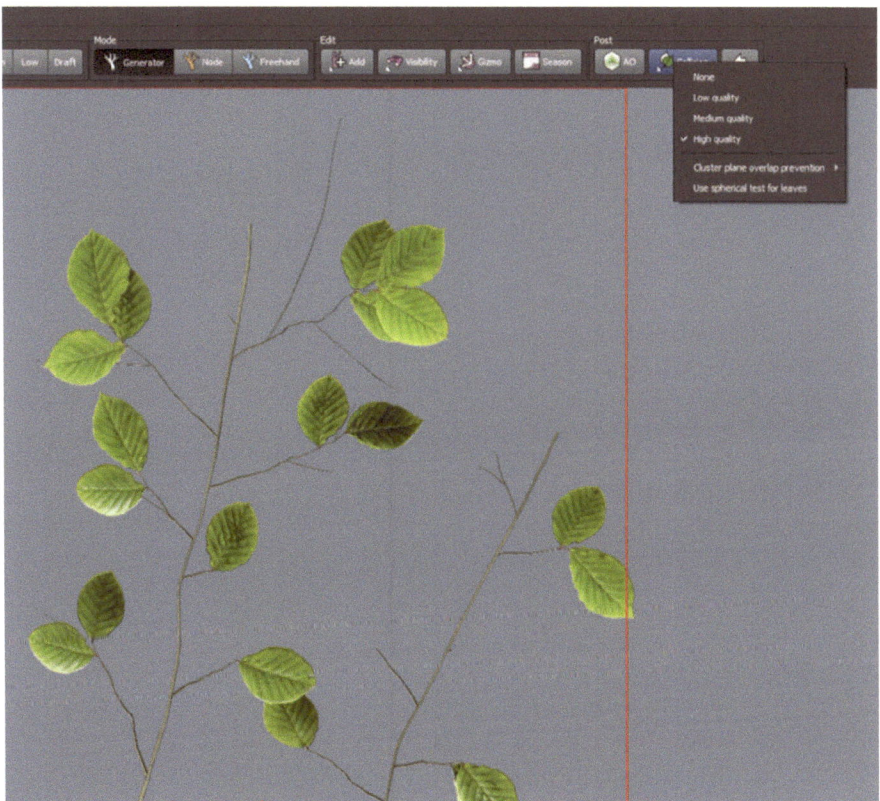

FIGURE 6.36 Example of the collision settings in SpeedTree.

Essentially, the collision settings remove overlapping leaves; in my case, I have set it to High Quality. Now, there could be cases where you want to change how the collision behaves, and this can be done per node. If you select your Leaf node and go to the collision tab, you will find a couple of settings in the drop-down menu.

- Everything: This will look at every other leaf node in your file and remove overlapping leaves.
- Ignore siblings: This will ignore leaves from within the same node and nodes connected to the same parent node, so if you have built something very

specific, you can turn this on to ensure it affects other leaf nodes but maintains any specific edits you have made to them.

- Knock out others only: This is similar to ignoring siblings, but it will collide with nodes connected to the same parent node.
- Nothing: self-explanatory.

With the collision settings turned on, you can either use the Art Director Gizmo or the Deformation settings in the Skin Tab to add some Folds, Curls, and Twists to the leaves. Check in with your reference to ensure you do this accordingly; remember to use the Variance Settings to introduce some randomness.

You can now copy over the Leaf node to the Little two branch node and the Big Branch node and tweak these according to your references. It has been pointed out before, but I recommend spending some time with each step to make sure you thoroughly understand what is going on, but also don't shy away from adding elements of your own and using settings not described in the chapter, the methods described are very generic, and I recommend aiming for a more detailed result than in the examples.

A great way to add more variety is to use decorations. You can find these under Add Geometry, Decorations. Adding these to your branches will go a long way toward breaking up the texture and adding some natural variety.

Refer to Figure 6.37 to get a rough idea of what your branch should look like before proceeding to the next step.

A good thing to consider in this step is how you will end up using and reusing these branches. Resources in a real-time environment are often sparse, and we must always consider how we optimize our content to be as efficient as possible.

On a texture sheet, you usually only have enough space for one large, a medium, and perhaps some smaller branches. On top of that, we also somehow must fit a lifecycle into the same texture sheet. There are many tricks to achieve this with limited resources.

For this reason, I always leave gaps between my branches. This allows me to cut them out and efficiently extract smaller branches from the big branch. Later, we can use these to create variety and volume in our branch clusters.

Refer to Figure 6.38.

The outlines suggest which parts of this branch can be reused, and later in this chapter, I will explain how to do this in SpeedTree using Anchor points.

We still need to create more branches of different sizes, but now is a good time to bake and see if it works on our tree. Setting up a pipeline for the whole setup as early as possible is beneficial for judging the whole product and seeing if this is going in the right direction.

To do this, ensure your whole branch is within the safe frame we set earlier, then press F10 to open the Export Material window. Set the Width and Height to 4096. We can leave the rest of the settings and press OK, it will then ask you for a location to save the textures, I saved mine here:..*CreatingHighQualityVegetation\workfiles\speedtree\ clusters*

Refer to Figure 6.39.

FIGURE 6.37   Example of the big branch with all leaf nodes setup.

This exports all the textures we set up from the perspective of the viewport camera, compressing our entire branch into a single 2D texture that we can use as a branch card in the tree.

Double-check the textures to see if there is any weirdness. Compare them to the ones in the source folder if you are unsure, I usually forget to set the Subsurface to black in the tiling texture; since this is a branch, we do not want it to let through any light, so keep an eye out for that.

Return to the SpeedTree file containing the tree if the textures look okay.

FIGURE 6.38   Example of the big branch and which elements can be reused.

APPLYING CLUSTERS TO THE TREE

The clusters are not entirely done yet, so this step is an in-between step to ensure we can make educated decisions when finishing up the clusters and building the branches we still need. Therefore, when doing these steps, it is important to keep it simple and not spend any time perfecting it, all we are after is a texture on a cutout mesh that we can

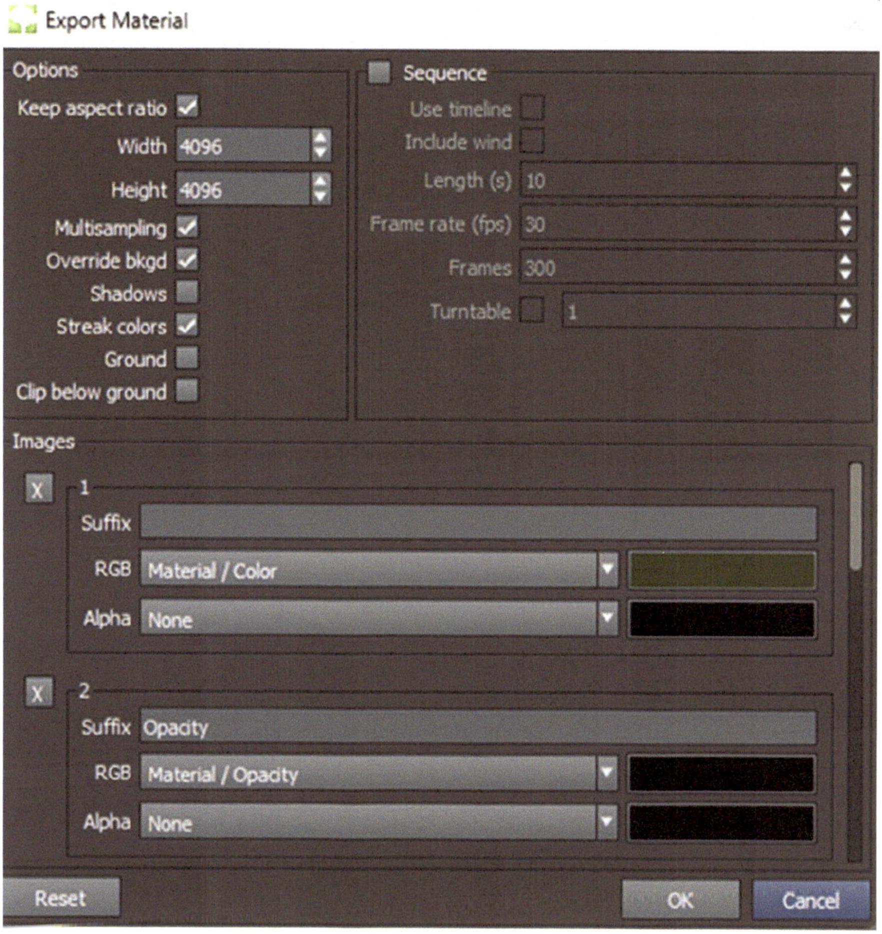

 Export Material

Options

Keep aspect ratio ☑

Width 4096

Height 4096

Multisampling ☑

Override bkgd ☑

Shadows ☐

Streak colors ☑

Ground ☐

Clip below ground ☐

Sequence

Use timeline ☐

Include wind ☐

Length (s) 10

Frame rate (fps) 30

Frames 300

Turntable ☐   1

Images

X 1

Suffix

RGB Material / Color

Alpha None

X 2

Suffix Opacity

RGB Material / Opacity

Alpha None

Reset   OK   Cancel

FIGURE 6.39 The export material window.

apply to our tree as early as possible so we get a good idea of what all components look like when used together.

Within the tree file, create a new material, call this Beech_Clusters or something similar, and apply the texture you exported from the ST_Beech_Clusters_01 file. You can find mine here:.../CreatingHighQualityVegetation/workfiles/speedtree/clusters/ST_Beech_Clusters_01.png

Since this is exported according to the SpeedTree naming convention, if you import the Color texture, it should ask you if it should automatically find the other textures; if you select Yes, it will automatically load in all different textures, so feel free to do so.

Navigate to the Meshes tab and create a new mesh. Click Edit Mesh, and in the Material section of the cutout editor, set the Material to Beech_Clusters or to whichever material you have assigned your cluster materials to.

Creating the geometry for the cluster is very similar to how we created geometry for the leaves earlier in this chapter. However, there is also an opportunity to work a little with masks. Since SpeedTree automatically generates geometry in between points, sometimes you create unwanted geometry, especially when working with clusters. Masks are a way to mitigate this issue and let SpeedTree know in which areas you do not want geometry to be created.

As usual, reset all default points and click around the mesh; for clusters, you should try to stick as close to the texture as possible. This is to reduce overdraw, similar to what is explained in Chapter 6; overdraw is the most expensive thing when it comes to vegetation, so it should be avoided where possible; one way to prevent it is making sure everything that uses an opacity map is cut out as close as possible.

Instead of having geometry go through the center vein, I recommend having geometry run over the main branch or primary branches to ensure that when this mesh gets deformed, it happens at a place where it makes sense texture-wise. I recommend not spending too much time on this on this as it is likely to change at a later stage, again, iteration here is key.

After your points are placed, or during placement, you can paint some masks to ensure geometry is only generated where you want it. Refer to Figure 6.40.

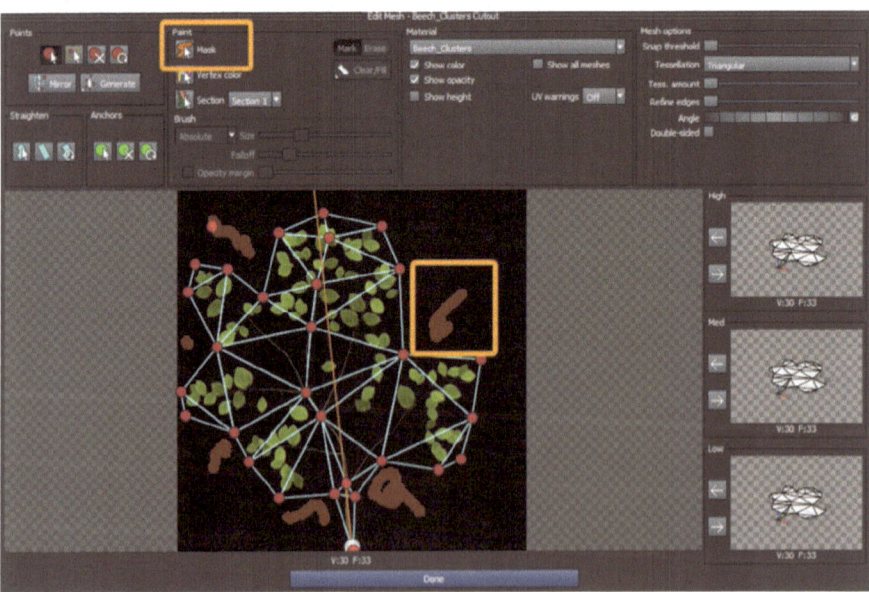

FIGURE 6.40   The cutout editor with masks highlighted.

You can be pretty loose with the mask painting. The goal is to highlight general areas where you do not want geometry created.

Lastly, go to the material tab and add the new cutout mesh to the Cutouts/Meshes tab.

With the geometry created, we can now add it to the branches. I recommend starting with the bottom branches, as these are somewhat isolated. This makes it much easier to see what is happening and how you want to set it up. We only want to add these clusters to secondary branches, so select those, Right-Click in the node graph view, Add Geometry, and select Leaves.

Worth double checking is the Show tab in the top left of the viewport, and make sure everything is visible; these settings are mapped to numerical shortcuts, and I find myself accidentally de-activating them and scratching my head as to why my leaves are not appearing. Refer to Figure 6.41.

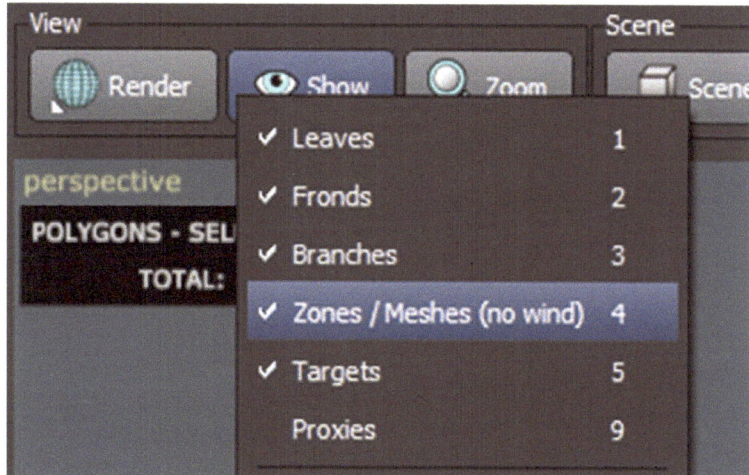

FIGURE 6.41   The show tab with all the different elements turned on.

A lot of elements in trees tend to repeat throughout trees. A leaf will eventually turn into a branch, and so forth; therefore, the rules for setting up our clusters are very similar to the leaves. Still, first, we must apply our material. Previously, we have done this by going to the material tab in the settings and selecting it from the drop-down menu, but there is another way that is, in some cases, more straightforward. You can click, hold, and drag the hand symbol next to the material onto the element you want the material to be applied to, in our case, our newly created leaves – highlighted in Figure 6.42.

With the material applied, start by scaling your clusters to a reasonable size according to the references. A value of 4 worked well for me. Then, in the Gen Tab, set the mode to phyllotaxy.

This should already give you a relatively flat structure, but the angle is a bit sharp, so in the orientation tab, use the align value to get them pointing slightly forward; check with your reference continuously to ensure you are doing this true to the source material. Sky influence can be beneficial here as well.

Depending on how dense you want to make the tree, you must play with the inter-node length to get a good result; I set mine to 4.

To create some variety, use Fold, Curl, and Twist deformations.

For now, this is enough to get a rough idea of where it is going. Right now, it is only spawning on the last generation of branches, leaving the tips of the secondary branches

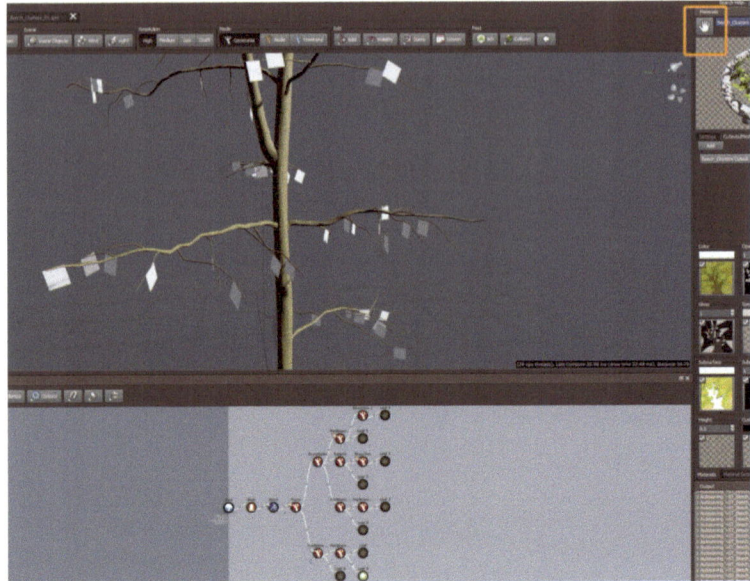

FIGURE 6.42   The hand symbol is highlighted.

empty. So, copy over the Leaf node and add it to the second-generation branches. You can then use the Internode length and Position in the Gen tab to push them toward the tips only.

Refer to Figure 6.43 for a good idea of what this step should look like.

For the time being, you can copy the same setup across the tree; feel free to make some changes, but keep in mind we will revisit this in a bit.

With the leaves copied across the tree, it is finally starting to look like a real tree, refer to Figure 6.44.

If you are comfortable, this is an excellent moment to import the tree into Unreal using Ctrl+E and export it as an.st file. This will be explained in more detail later, but the reason why we export as an.st and not.fbx, which is the industry standard export format, is that this way, we can inherit the settings we have set up in SpeedTree more easily, so the textures and shader will be partially set up and we just need to clean it up to fit with our preferences.

The benefit of getting the setup in Unreal is that you will be able to have a look at it in the context of your target engine, allowing you to make more educated decisions but also look at the tree under a different light and perspective which always helps to judge what is working and what needs to be fixed.

If you are unaware of how to do that, continue with the chapter, and we will get to it later.

As of right now, the most significant thing it is lacking is variety and volume, so open up the cluster SpeedTree file again to make more branches and finalize the branch we made before.

Within the Cluster File, we first need to set up a new base for our second branch. To do so, you can copy over the trunk and the first branch, then select the Trunk 2 node that

FIGURE 6.43 Speedtree viewport with the initial branch structure setup.

is created and use shortcut W to open the move gizmo and move it out of the way into some empty space within the safe frame. You can refer to Figure 6.45.

Keep in mind that depending on how your branch looks, it could make more sense to move it elsewhere; we will create a medium branch, so find some space suitable for that size.

When working on variations of things already built before, the power of SpeedTree starts to work in our favor, as most of what we need to do now is copy over nodes from the large branch and maneuver these into place on the medium branch. I again recommend looking at references and not diving into too many details just yet; the goal should always be to get these individual elements into the main tree as quickly as possible and then start refining from there.

After placing your trunk into some empty space, copy over the Big branch node from the large branch, set the Generation Mode to Phyllotaxy in the Gen tab, then set the Arrangement Style to Alternating (distichous). Use the Internode Length setting to

FIGURE 6.44   Tree with an initial canopy setup.

get a good number of branches and consult the references to determine this. In my case, an Internode Length of 0.3 worked well, play around with the Start Angle and then used node edits to get branches into the correct positions.

With the secondary branches in place, we can add some noise to the main branch and use freehand to get closer to references. Refer to step 1 in Figure 6.46.

1. The medium cluster without leaves
2. The medium cluster with tertiary branches and leaves applied.

Figure 6.46 also shows the next step, adding tertiary branches and leaves. To do so, we can yet again copy over the majority from the large branch and use the power of SpeedTree to move this forward efficiently. Double-check your reference to ensure you are on track, change the length and number of branch breaks, and adjust the start angle accordingly.

For the leaves, use the workflow explained earlier in this chapter, overshooting the leaf count a bit and then using node edits and collision to make them look natural.

An additional setting worth looking at is the Up Rotation. Using the Variance settings for the Up Rotation in the Orientation tab can help with some intersecting

FIGURE 6.45 Branch copied over to space in the texture.

leaves. Additionally, you can use the Art Director Gizmo to do this by hand if you need more control. I am using a mix of both. Refer to step 2 in Figure 6.46 to see the result.

The addition of other leaves could affect the collision's effect across the scene, so make sure to keep an eye on the main branch and adjust if necessary.

Repeat this process for a small cluster as well. Make sure to consult your references when doing so.

Another element currently missing from our texture set is bare branches. These help a great deal in introducing the tree's lifecycle and should never be forgotten. They can also help fine-tune the density of the canopy.

We can create these by copying over the whole structure of the small and medium branches just created, removing the leaves, and making them look slightly different than their respective counterpart. There is a quick way to do so in SpeedTree; if you

FIGURE 6.46   The medium branch with and without leaves.

Right-Click a node, you can select "Randomize Selected." This helps a lot when creating variety. If you do not get the desired result immediately, click it a couple of times and see if something sticks; remember that this can also get you eighty percent there, and the last twenty needs to be done with node edits.

Refer to Figure 6.47 to see the whole canopy kit created.

FIGURE 6.47   The completed canopy kit.

Export this using F10 to open the Material Exporter and overwrite the textures created for our test earlier in this chapter. Then, open your tree file, navigate, and refresh the textures if they haven't done so automatically.

We can edit our previously created cutout mesh to fit the new textures, but for all other textures, new meshes need to be made in the Mesh tab, re-assigned to the Material, and then used the Cutout Editor to create meshes for the rest of the added branches. I created one large mesh, which will do all the heavy lifting, three medium branches, two small, and two dead branches.

I cut out two of the three medium branches as extracts from the large branch, as explained when we were building the larger branch. Refer to Figure 6.48 for a refresher and an example of how this would look in SpeedTree.

FIGURE 6.48   A medium branch cut out from the large texture.

WORKING WITH ANCHOR POINTS

In order to combine all our branches together into a nice volumetric cluster we can use Anchor Points.

Anchor points are points created in the cutout editor that will guide the placement and direction of additional leaves. This will allow us to plug a leaf node into another leaf node and give us control over where the second layer of leaves will be placed.

Start by opening up the large leaf mesh in the cutout editor. You will find the Anchors section at the top left in the cut-out editor. Highlighted in Figure 6.49.

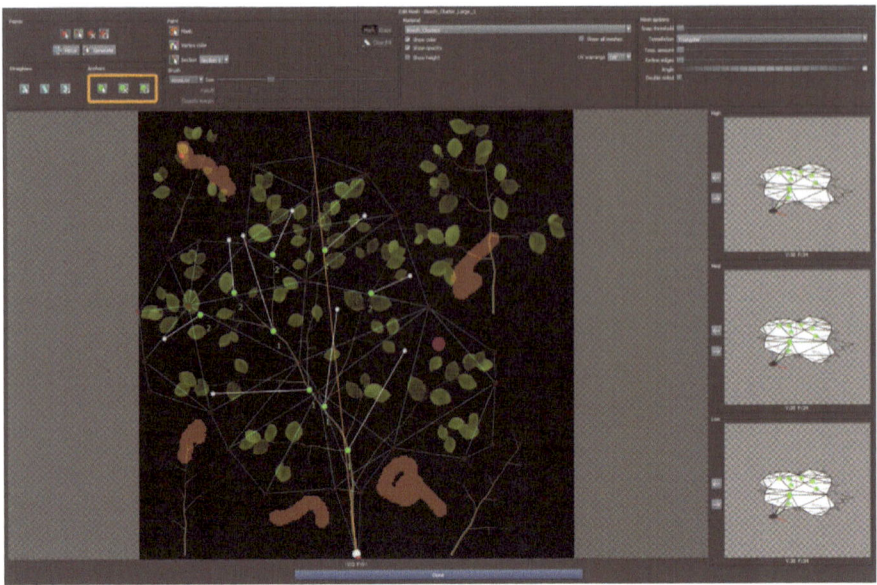

FIGURE 6.49   The cutout editor with the anchor points highlighted and A medium branch cut out from the large texture.

You will also see the anchor points I placed in Figure 6.49. To place them, use the left button and click on the location where you want new leaf branches to grow out. I suggest picking locations where it would happen naturally. It could be that you must thin out your large branch to accommodate the added density. If you click, you will see the Anchor point pivot in green and a white dot connected by a line.

This white dot indicates the direction of growth, so make sure the branches point outwards and do not overlap too much with other branches. We can also give anchor points an ID and use that ID later to control which of our branches can spawn on which ID. In Figure 6.49, you will notice that I have set two IDs, 1 will be used for medium-sized branches and placed toward the start of the branch, and 2 will be used for our smaller branches and, therefore, placed toward the end.

To set an ID, you can Right-Click anchor points, select ID in the drop-down menu, and then select which ID you want to assign. You can assign up to five IDs, but I recommend keeping it simple as you can quickly lose track of what is happening, especially if you are new to the SpeedTree workflow.

Once you have placed some anchor points, make sure you also use the arrows on the right to load them into the mesh slots. If you are unsure what to do with them, remember that you can always change and adjust them. I recommend placing one anchor point until you understand the workflow and gradually adding in more.

I usually want to see exactly what happens when working with anchor points, so I will initially focus on one branch before I apply it to the whole canopy.

I think it is easiest to see what we're doing at the bottom branches, as they are already isolated, so I went ahead and called that Leaf Generator "LargeLeaf" and copied that over two times, naming one of them Medium and the other SmallLeaf you can see what that looks like in Figure 6.50.

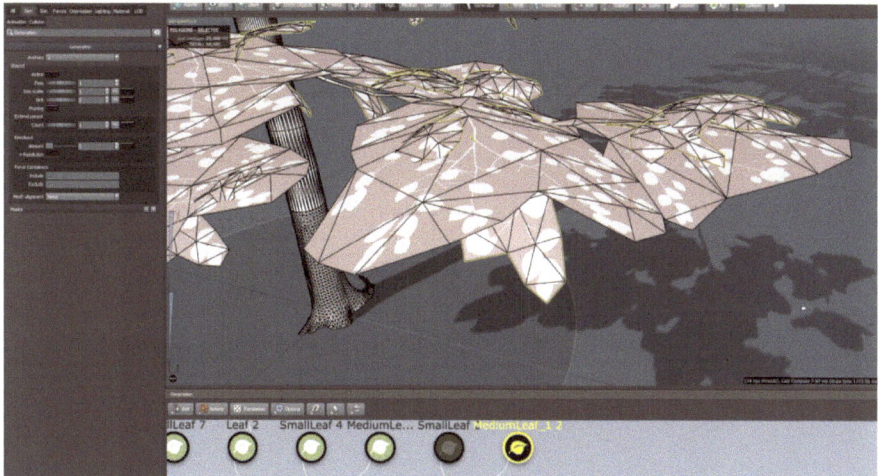

FIGURE 6.50   Leaf chain visible and the result is visible in the viewport.

I have set the viewport to Wireframe mode as it is easier to see different elements.

You can start by copying the LargeLeaf setup and plugging it into itself, but you will find it only spawns the Large Mesh on top of it. To prevent this from happening, you need to specify which meshes are allowed to spawn within the Node Editor. Earlier, we made three different materials for small, medium, and large leaves, but this is an excellent opportunity to show you another way.

Navigate to the Material Tab of your Medium Leaf Node, and under type, you will see a Material Slot and a Mesh Slot. Start by changing the mesh in the Mesh Slot to a medium-mesh, then use the + arrow in the top right of this segment to create three more mesh inputs; set the material to your cluster material, in my case, that is called Beech_Clusters and then set the Mesh to the other Medium meshes, if you have made more or less than three medium meshes you need to adjust this number accordingly. Refer to Figure 6.51 to see the setup I used.

Additionally, you will see that I have set the weight of the Dead Medium branch to 0.5, as I want most of the branch to be alive. The weight setting determines a branch's spawn chance, so it is recommended to play around with these values to get the result you want.

You can then shape the leaves using the Deformation tab, make sure something is visible from all angles, and consult your reference to see what makes sense; Beech clusters tend to grow very flat, so do not go overboard with these settings.

You can then copy over the Medium Leaf Node and call it SmallLeaf, then also plug it into the Large Leaf node, navigate to the Material tab of this node and shift the medium meshes for the small meshes.

You might notice at this point that both the Medium and Small leaves are spawning from the same location, and this is why we have set our IDs earlier in the cut-out editor.

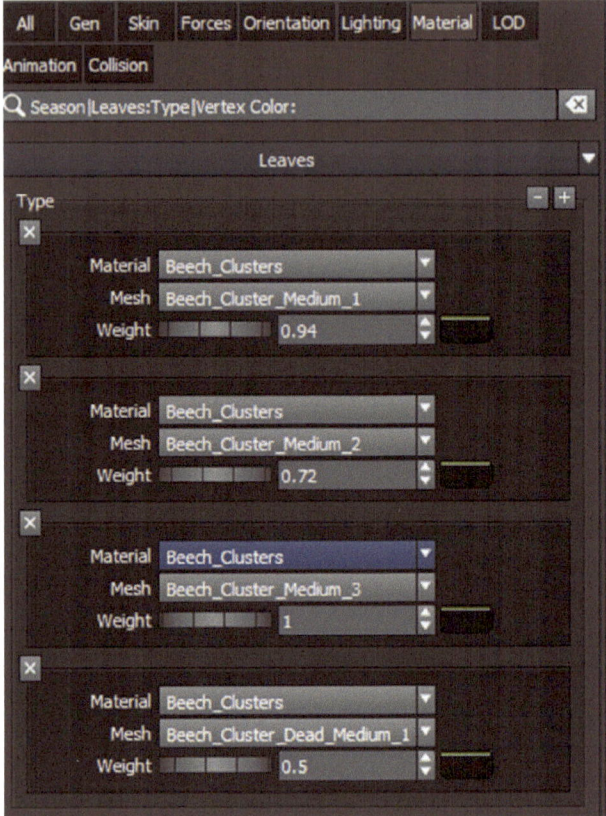

FIGURE 6.51   A material setup to show only medium leaves.

To make the small leaves spawn on ID 2 navigate to the Gen tab and set the Anchors setting to 2.

This concludes the technical setup for the anchor points, so at this step, I recommend playing around with it until you are happy with how it looks visually, and then copy over these nodes to the rest of the canopy structure to apply it throughout the tree. Refer to Figure 6.52 to get an idea of what this should look like.

At this point, we have the whole SpeedTree pipeline set up, and I encourage you to play around with this setup. Make a copy of your original file if you are worried about losing any data; something not communicated in this book is the number of iterations a single asset goes through; you only see the final or close-to-final images. To give you a rough idea, I go through about 100–200 iterations per tree on average, sometimes tweaking the shape of the trunk, adjusting significant positions of the secondary branches, and sometimes changing the direction of one leaf. It is good practice to try as much as possible to make substantial changes early on, and make your way to smaller and smaller iterations, but in reality, making more significant changes late in the production of an asset is often unavoidable, so do not shy away from making this a continuous habit, a common thing said throughout the industry is that the first eighty percent of an asset

FIGURE 6.52   A branch cluster with smaller branches placed on anchor points.

takes shorter than the last twenty percent, this chapter should have gotten you through the first eighty, and the final twenty is up to you.

Refer to Figure 6.53 to see at what point I considered my version to be ready for export and further tweaking in Unreal Engine

FIGURE 6.53   Beech_L_01 in its entirety.

When looking at your tree in SpeedTree, there is a couple of things I recommend paying attention to:

Make sure your leaves are somewhat uniformly scaled. Too large or small leaves or a mix all over the tree are a quick way to make your tree look fake, but with so many settings to keep check of it is very easy to have this happen to you, so have a look over your tree to make sure everything is looking uniformly scaled, I often find myself scaling the leaves at the top slightly larger this allows you to fill up the canopy quicker while using fewer cards, reducing overdraw and triangle count, do not go overboard this though, I usually scale them about ten to fifteen percent larger – performance is however not a significant concern for this project. Therefore, if you want to keep your leaf size the same across the whole asset, feel free to do so.

Secondly, under the Render Tab, set your view to Scribed and look at what the wireframe looks like. The goal is also to have a somewhat uniform look to this. You can opt for having more geometry toward the bottom of the tree as this is closer to the player camera, and gradually reduce it as you go to the top; try to avoid having pitch-black wireframes when looking at it from a reasonable distance. But keep enough geometry to get nice and smooth deformations and displacement; it can be tricky to find a balance, and sometimes it is a trade-off – I find the curves work well when optimizing the tree, there is more information on optimization at the end of this chapter as these allow you to have more geometry relative to the length of the branch.

What you see in Figure 6.54 is the Radial Segment curve of one of the bottom branches. The Absolute value is set to 8, and the relative value is set to 2, which means that to a certain degree, it looks at its parent to decide how many radial segments it gets, but it will always have at least 8. The curve then controls this, which is set to Linear Decay. You can see that at the start of the branch, there are more radial segments, and as it grows longer and smaller, therefore less visible, the radial segments get reduced.

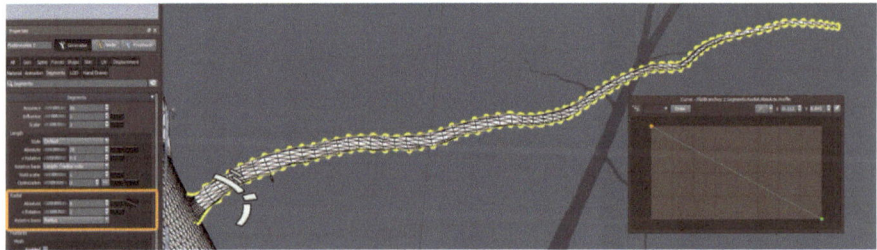

FIGURE 6.54 The radial segments graph with a gradual decrease.

Aside from the technical aspect, the tree must also look good visually. In many cases when I see work from inexperienced artists the trunk and branches that come out of SpeedTree look decent, but the canopy could be better, a common mistake is making a canopy overly dense. Generally, density comes from stacking multiple layers of branches on top of each other or various trees in the case of a forest, but each branch is usually a lot sparser than you think. I recommend looking at your tree from the angle it is most viewed at, in our case from the bottom, and seeing if you can see some negative space and holes in the canopy. If it turns out to be a solid shape, you have made it too dense, and it will not look visually pleasing.

It is a delicate balance to strike and will most likely require some iteration. The best I can describe is ensuring your canopy has room to breathe. In my experience, I must always take a step back and iterate on the canopy texture to remove some leaves, so I do not shy away from going back and forth between the entire pipeline to get a satisfying result.

This process can take up to a couple of hours.

Of course, the best way to judge what needs to be changed is to get the trees into our target platform, Unreal Engine. So, if you have something ready to be exported, I advise you to continue along with the chapter and make these decisions based on what you are seeing in Unreal Engine.

THE WIND WIZARD

Once you are happy with how your tree looks in SpeedTree, we can run the Wind Wizard to ensure some movement when the tree is in Unreal. To do so, open the Wind tab at the top of the viewport and open the Wind Wizard; refer to Figure 6.55.

The wind wizard will set certain values for your tree and is relatively self-explanatory. However, for a simple breakdown, refer to Figure 6.56.

There are three significant components to the wind wizard:

1. General: In this segment, you set the type of tree. In our case, we can use a Stately Shade Tree; other presets include bushes, grasses, ferns, or palms. You can run the Wind Wizard as often as you like, so feel free to try out these settings to get a rough idea of what they do. Experimentation is the way to understand things.
2. The type of geometry it is dealing with. The obvious choice here would be to pick the Small, Rustly preset, but since we use a card workflow, the Wide Flat fronds worked better. I suggest trying both and seeing which one you like best.
3. For each Leaf node, you can set whether you have used it to build Leaves or Needles, fruits, or flowers or if it is a stationary object. There are use cases for all of them, but in our case, we can set all of these to Leaves/Needles.

When you have set up the Wind Wizard, you can press OK. If the Wind did not automatically activate, you can do so in the Wind Tab or press 8.

Lastly, before we export, we need to bake the texture information from our Bake Stitch, have a quick look over to see if it is blending correctly, then switch to Node editing mode and select the parent node (in our case, the node that holds the photogrammetry trunk) then in the settings go to the Bake Stitch Tools tab and select the Assign New Material button. Refer to Figure 6.57.

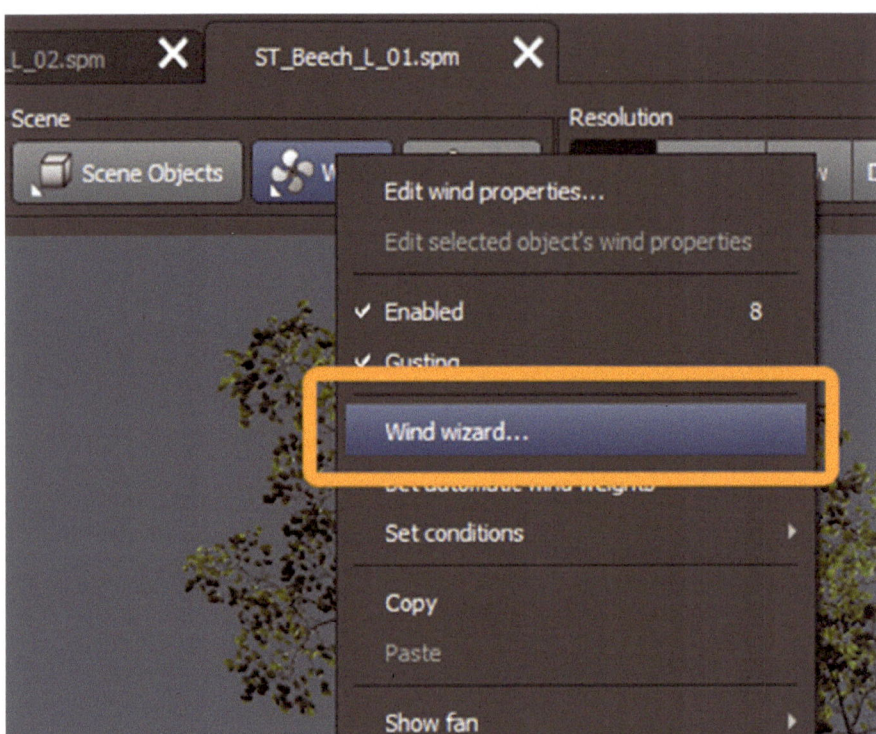

FIGURE 6.55   The location of the wind wizard.

This will create a new trunk texture with the tiling material blended in. It should be automatically updated, but if you see any errors, double-check that the right material is applied; this applies to SpeedTree and Unreal Engine.

EXPORT TO UNREAL

SpeedTree has an excellent pipeline for getting our tree into Unreal, and it does a lot of the setup for you; however, it is essential to understand what is happening and why.

To export your tree, go to File, Export Mesh, or press Ctrl+E. This will open the Export Window.

In this Window, under Preset, you can select UnrealEngine(ST9), but there are a couple of changes we need to make. Under Grouping, it is set to export all LODs, and since we have not set up any LODs, this will get in our way if we leave it, so set this to Highest Only. If you have built LODs, you can keep it default.

FIGURE 6.56   The wind wizard.

By default, the Atlas Segment is set to Non-Wrapping, which means every texture that is not a tiling will be packed. This will put our canopy and photogrammetry trunk textures on the same texture sheet, so in this case, we want to opt-out and select Nothing.

The rest of the settings are set up correctly; we can leave those as they are. We can save our setup as an export preset to make re-exporting easier. To do so, press the three dots next to the Preset, select Save New Preset, and give it a name. Refer to Figure 6.58 for the correct Export setup and where to save this as a preset.

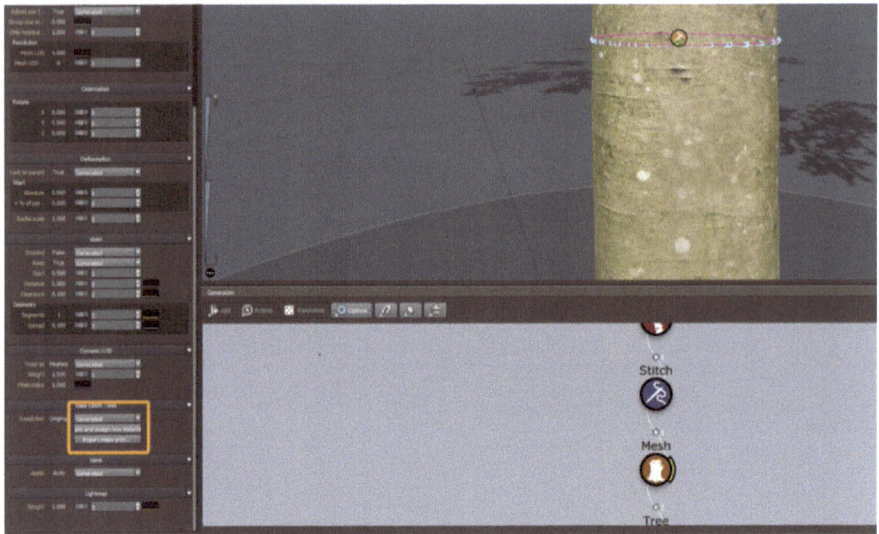

FIGURE 6.57   The bake stitch button.

Press export and set an export location. I am exporting here using the file name as the export name..*CreatingHighQualityVegetation\\workfiles\\speedtree\\export*

After the tree has been exported, open the Unreal Project and make a new Trees folder in the Vegetation folder. For reference, mine are saved here: ...\\ *CreatingHighQualityVegetation\\workfiles\\ue\\CreatingVegetation\\Content\\Creating Vegetation\\Assets\\Vegetation\\Trees\\ST_Beech_L_01*

Important to note is that when importing directly from SpeedTree, it will create the materials and textures for you based on the names you gave them in SpeedTree. Therefore, I have done some cleanup afterward and have also created a Materials and Textures folder. Initially, I imported everything in the Trees folder and then moved the imported elements according to the folder structure.

Feel free to do the same or keep it all in the Trees folder. For personal projects, the way you set up your folders is heavily based on personal preference, and there is no technical dependency on having the same structure as described in the book.

In the Trees folder, Right-Click and select Import To, similar to how we imported the plants in Chapter 6, browse to your SpeedTree export folder, and select the .st9 file you want to import. This can take a little while to load, but it should import your mesh, textures and materials for you, refer to Figure 6.59.

You can drag this into a level to have a look at how it looks in-game; at this stage, it is very typical to notice things you have not seen before in SpeedTree; things respond slightly differently, and that can highlight some issues or happy accidents, that you weren't aware of before. And for me, this is where the fun starts.

With the tree visible and being able to virtually walk around it and see it in the target engine, we can get into the iteration process. The iteration process is dependent on how your tree looks and how you want it to look, so I have not described my iterations in a lot of detail in this book, but to give you a general idea. The iteration process can take

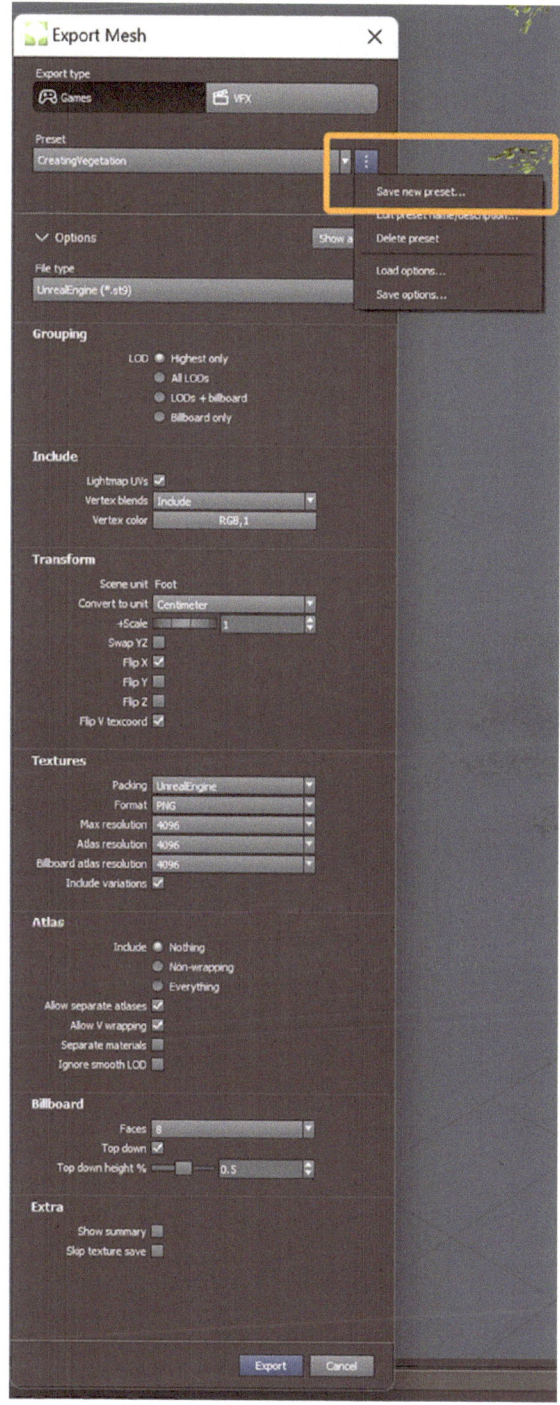

FIGURE 6.58   The SpeedTree export window.

FIGURE 6.59   The content browser with our tree imported.

up to twenty hours with some trees, so do not shy away from going through a couple of iterations and tweaking things as you see fit. This process is highly individual per tree, and it depends on choices made before, but there are some general things to look out for:

Does the canopy have enough breathing space? Look at your canopy from its primary viewing angle and see if there is enough negative space. Refer to Figure 6.60.

FIGURE 6.60   Example of good negative space and poor negative space.

On the left, you see an example of enough negative space; the leaves are individually readable, you can see clusters and depth, and it allows light to pass through, compared to the right, where the whole canopy turns into a solid wall. This is unwanted and will become especially problematic when the tree is used in the context of a scene if you find yourself with a too dense canopy, try and copy over the tree a couple of times. You will notice that they will stop all light, resulting in bland shadows.

To fix this, I recommend decreasing the amount of leaves in both the tree and the leave clusters; the problem is either too many cards spawned in SpeedTree, or the clusters textures are too dense, so for example, increase the internode length a little in the tree and then remove some leaves from the texture and repeat this process until you are happy with the density. It will take a couple of iterations to strike a good balance.

Additionally, a common problem is not having enough geometry to support all the deformations or not having visible harder corners, as displayed in Figure 6.61.

At the top, you see a nicely deformed branch. There is enough geometry to support all the shapes, and no harsh corners are visible. Compared to the bottom example, if you

FIGURE 6.61   Example of lacking geometry.

notice shapes like these close to the player camera, they should be fixed by going into the Segments tab and increasing the amounts. Those harsh shapes quickly make your tree look cheap and detract from the visual fidelity.

There are times in game production when this is unavoidable due to performance constraints. If that is the case, I would recommend shifting the geometry toward the area of the tree most likely visible by the player and having the cheaper geometry in places that are less visible or visible from further away.

Earlier in this chapter, we also discussed constantly rotating around your tree. There is a common issue that can occur if you look at your tree from only one angle: if you have multiple branches overlapping each other, they run the risk of creating one thicker, straight, and somewhat dull-looking shape. I suggest keeping an eye out for this issue and, if it happens, bending some branches to split up the branches; that way, your tree will look more attractive from all angles. Refer to Figure 6.62 to see what this issue looks like and how to fix it.

Lastly make sure to check for branches that appear to be floating, refer to Figure 6.63 for an example. The reason this happens is that the stem that it is attached to is too thin and therefore turns almost invisible, you can either opt for making your branches a bit thicker or push them in a bit, so they are hidden in the canopy.

Alternatively, you can opt for removing the whole branch, it might require some back and forth in SpeedTree and Unreal to get to a desired result.

Aside from these common issues, this is an excellent time to review your own work and make changes where you see fit. Sometimes, it helps to take a small break and return with fresh eyes.

Once you are happy with the large tree, it is time to use your newfound or sharpened knowledge and build some variations; you can reuse the elements we have already

FIGURE 6.62   Example of bland shaping due to overlapping branches.

FIGURE 6.63   Example of floating branches due to thin stems sticking out.

created, and if you want, use some of the other photogrammetry trunks provided with the sample files here: *...\CreatingHighQualityVegetation\workfiles\blender\export\Phototrunks* to create different sizes as well.

In the next chapter, we will focus on building a portfolio-ready scene, and I recommend building at least a Large, Medium, Small, and shrub-like asset before continuing. This will help solidify the learnings from this chapter and make your portfolio piece a lot stronger. If you want to continue without building more trees, you can also use the ones I created. They can be found here: *..\CreatingHighQualityVegetation\workfiles\ speedtree\export*

Refer to Figure 6.64 for an example of a tree line-up before moving on to scene building.

FIGURE 6.64   The complete tree lineup.

With our tree assets lined up, we can move on to the next step, building a scene.

Remember that when building the scene, it is still recommended that you make changes to your trees and plants as you see fit. This will be the first time we see all our assets come together, and it is an excellent stage to review our work if it is being used together. In fact, I always strive to get to this step as quickly as possible, as I believe seeing the assets in context is the best way to decide what is essential and what is not.

OPTIMIZATION

Once you have your whole tree line-up ready it is a good idea to make some optimizations, especially if you are doing this for a game environment.

I recommend turning on Wireframe mode and looking for densely packed geometry to see if any optimizations can be made. If you do find yourself with an extensive number of triangles, the best way to optimize is in the Segment Tab of SpeedTree.

I usually remove any Relative segments and use the Absolute sliders instead. Then, I use the curves with a Linear Decay going to have fewer segments toward the end of branches and higher up in the tree. A good number to aim for is 100,000 triangles for the largest trees and 20,000 for the smallest, but this depends on what size trees you have made, so it is hard to give good estimates.

If you are building the tree solely for a Nanite workflow in Unreal Engine. In that case, it is best to add as much geometry as needed to remove any opacity, as it is much more expensive than triangles.

Setting up a Wind Material

7

VERTEX COLORS EXPLAINED

We will set up two things for the wind material: a global gradient from the bottom to the top of the asset based on the bounding box to control the general swaying and on top of this, we will layer smaller-scale movement using three vertex colors: red, green, and blue.

In case you get stuck on the technical side of things you can always refer to the example file here: ..\CRC\CreatingHighQualityVegetation\workfiles\ue\CreatingVegetation\Content\CreatingVegetation\Materials\MaterialFunctions\MF_Wind.uasset

A basic understanding of vertex color is required to follow along with the chapter, so let's start by figuring out what vertex colors are:

Essentially, every vertex on a mesh contains data. Vertex colors are part of this data and contain a number from zero to one in each channel (RGBA). This data can then be used within game engines to drive certain effects.

In the case of wind, we can, for example, define that for the red channel, we apply a value of 0 at the stems and have a gradient toward a value of 1 at the tip of the leaf. We can then tell our wind material to look at this gradient and say we want you to have no effect at 0 and maximum effect at 1. Doing this just for the Red channel will generate very simple results, but as soon as we start layering multiple vertex colors with different effects, this data set becomes one of the Vegetation Artists' most powerful tools.

In our end, we will set up a primary and a secondary wind, the primary wind will be controlled by a gradient based on the Pre-Skinned Local Position of an asset, offset by the Blue Vertex Color channel and the secondary wind will be controlled by the SimpleGrassWind node offset by the Red and Green Vertex Color channels.

As explained when building the assets, the blue channel is a random value for each cluster that we want to bend individually, this will offset the overall swaying of the plant per cluster, resulting in a more natural-looking sway, instead of the whole plant swaying, we can see some differences within the pant.

The red channel will control where we want and do not want the secondary wind to be applied; we like to have this effect be stronger at the tip of leaves and be completely gone at the stems.

DOI: 10.1201/9781003492283-7

And lastly, the green channel will control the offset of the secondary wind. If we leave it at the default settings, the whole plant will get a uniform offset, but in reality, not every leaf moves in the exact same way, and not every leaf moves at the same time, using the green channel to offset this will create a more realistic look and feel to the wind.

SECONDARY WIND MOVEMENT

Shaders can be overwhelming, and wind shaders are no different. You can make these as complicated as you want. Usually, this responsibility leans more toward that of a technical artist, but as a Vegetation Artist, it can become a bit of a shared responsibility. It is a big plus for any Vegetation Artist if you are technically capable, so I recommend trying to learn as much as possible about the technical parts of creating game assets.

We will keep things relatively simple for our material, using only basic math and built-in functionality. We will start with just getting things moving around and expand on it from there.

Start by opening the M_VegetationMaster material created in Chapter 7 and putting down a "SimpleGrassWind" Node; this node gives our foliage a basic wind operator and has four inputs: WindIntensity(S), WindWeight(S), Windspeed(S), and AdditionalWPO(V3). The (S) stands for scalar, which means it will only accept single-value inputs, like the Constant1 we have been using earlier.

We will not be using the AdditionalWPO, but it does require something to be put in not to throw an error, so for starters, put down a Constant node with a value of 0, plug this into the AdditionalWPO, and create another Constant, convert this to a parameter and call it WindIntensity, use a default value of 0.1 for now and plug this into the WindIntensity, we can also make a new group and call it Wind, that way we ensure all our Wind settings end up in the same place when we create a Material Instance.

For the WindWeight and the WindSpeed, we will be using the vertex colors set up in all our assets, and now we will start seeing why that was important. As mentioned before, the red channel defines where the wind will have an effect and where the wind will not have an effect, which is precisely what the WindWeight controls, to sample VertexColors in our material, we need to put down a VertexColor node, Right-Click on the grid and search for VertexColor, plug the Red Channel of the node into the WindWeight Input.

We have also prepared our assets with a random green vertex color for each strand; we will use this for the WindSpeed, offsetting the speed a little bit per element to generate a more believable result. Put down a Multiply node and plug the green channel of the Vertex Color node into the A slot; create a Scalar Parameter by holding down the S key and clicking on the Grid; this is the same as creating a constant node and converting it to a parameter but a little bit quicker. Call the parameter Wind Wiggle Speed, set the default value to 0.2, add it to the Wind Group, and plug this into the B slot of the Multiply node.

Create another scalar parameter and call this one "Global Wind Speed" with a default value of 1; for the time being, this will not do anything. However, we will reuse this parameter in other parts of the wind material to control both the primary and secondary wind animations.

We must add this logic together to start offsetting the wind, so create an Add node and plug in the Multiply. Then we will create another Multiply node, plug the Add node into the A slot and the Global Wind Speed Parameter into the B slot, and then plug this whole node network into the WindSpeed of the SimpleGrassWind.

To summarize what we have done: we have multiplied the value we have in our green vertex color by 0.2 and add this on top of itself, the add node will ensure lower green values are also affected, in the event of pure black (or 0) values that can't be multiplied as 0.2 multiplied by 0 will remain 0. Then at the end we also multiply this logic by a global Wind Speed Plants parameter, doing this ensures we have one global wind setting to rule them all.

The complete setup can be seen in Figure 7.1.

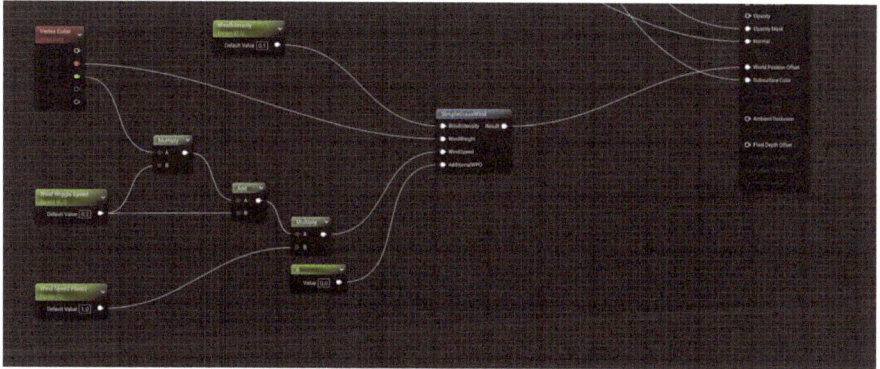

FIGURE 7.1 The SimpleGrassWind node with our initial WindSpeed and WindIntensity setup.

ADDING WIND GUSTS

In order to add some variation to our secondary wind, we can add some randomness to how fast and strong the wind moves simulating gust. You can see the whole logic in Figure 7.2.

There are a couple of new things in here, so let's go over the setup, if you need to you can reference the example file while reading through the chapter: ..\ *CRC\CreatingHighQualityVegetation\workfiles\ue\CreatingVegetation\Content\ CreatingVegetation\Materials\M_VegetationMaster.uasset*

Place down an Absolute World Position node, if you are interested to see what this does, you can plug it into an emissive slot of the texture, and you will notice a lot of colors, but what we are after is a grayscale image, essentially trying to fake a gust of wind going through the world, using the Absolute World Position for that will make sure that plants in a similar world position will respond to the same gusts of wind. We do not need

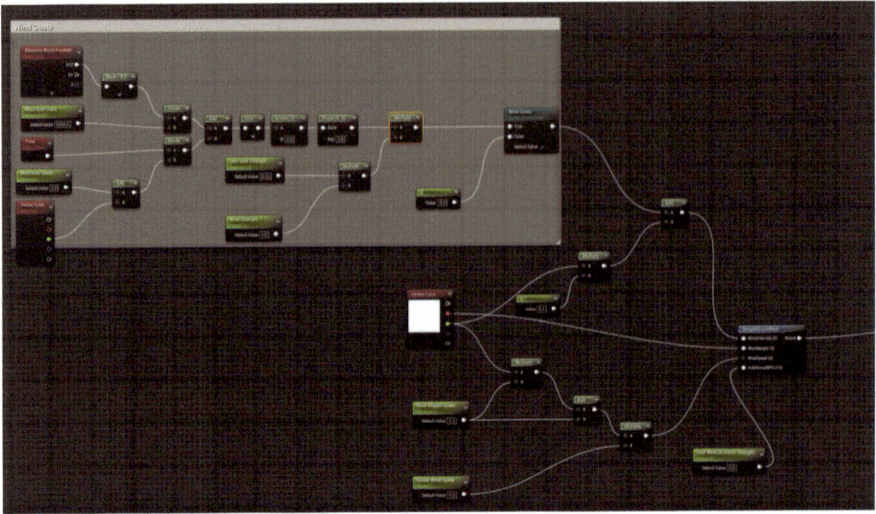

FIGURE 7.2 Node graph showing the wind gust logic and where to add this in our secondary wind.

all the colors but just a gray scale. To achieve this, plug it into a Mask node and isolate the R channel, divide this by a Scalar Parameter with a default value of 5000, and plug this Divide into the emissive slot then I recommend playing around with the wind gust scale to see what this does.

To create a bit more variety, create a Time node, this will allow us to add the passage of time to the graph, plug the time into a Divide A slot, and create a Scalar Parameter called "Wind Gust Speed" with a default value of 2 and add this node to the green channel of a Vertex Color Node, this to ensure the value of 2 gets randomized by what we have put in the mesh vertex colors, we are using an Add in the event the mesh has no vertex colors. Explained in more detail earlier in this chapter, then plug this into the Divide B slot.

We now have 2 Divide nodes, one from the World Position and one from the Time; add these together using an Add node and plug this into a Sine node to ensure the gusts move between a minus one and one range.

Divide the sine by 2; you will notice that in Figure 7.3, I have not used an actual parameter for this; this is because I do not need control over this in the material instance; you can do this by just selecting the node and typing the value in the Details panel of that node. This way we can keep our node grap looking a bit cleaner.

Again, you can plug this logic into the Emissive channel at any time to visualize what is happening; even better if you put a couple of assets down in the scene to show the effect over multiple assets. You might notice little contrast, which makes the effect vague. We will run it through a Power node with a value of 2 to fix this.

As per usual, we want individual controls and a global controller, so put down two scalar parameters: Leaf Gust Strength with a default value of 0.15 and Wind Strength with a value of 1. You can either reuse a previously created Wind Strength node or make sure that the name is identical; that way, your graph looks a bit cleaner because you

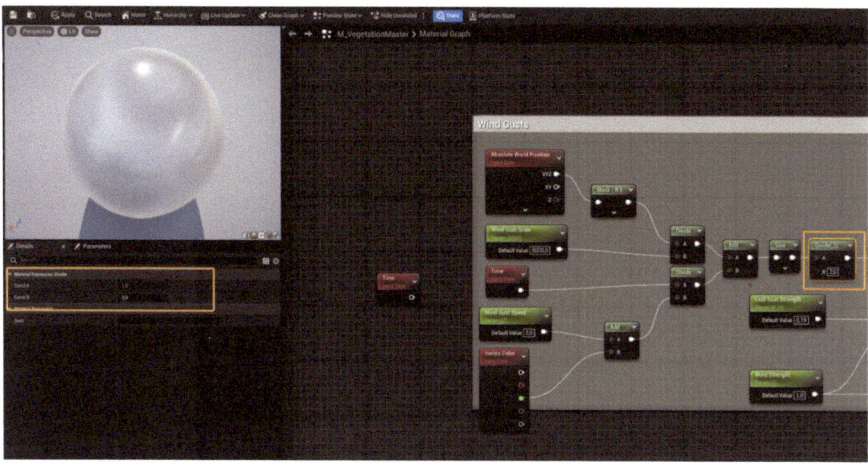

FIGURE 7.3 Selecting the divide node and adding the values in the details tab.

avoid crossing wires and since the name is identical, Unreal will automatically recognize and reuse the value put elsewhere in the graph. Do keep in mind that if you ever change the name, you will need to do this in every place where it is used, so to reduce chances for human error, it is better to have one node and plug this in in multiple places, but if you want the whole thing to look cleaner using multiple nodes is the way to go, feel free to choose whichever you are most comfortable with.

Multiply both parameters together, then put down another multiply node, plug the Power node into A and the multiplied parameters into B. In order to keep our graph nice and tidy, let's have a look at switches. If you search for StaticSwitchParameter and place one in the graph, I called mine "Wind Gusts" you can plug the Wind Gust logic into "True" and plug a Constant1 with a value of 0 into False and set the default value to true. You can add this switch to our Wind group as well. This will expose a button in the material instance where we can turn this logic on and off per instance. This means we can make our setup a bit cheaper as we do not need the wind gusts on every asset but only on the ones where it will be visible.

If you have everything plugged in you will notice that every once in a while the wind will move a bit faster and stronger, that is the effect of the wind gusts we have just added.

PRIMARY WIND MOVEMENT

We started with the secondary wind because the setup is a little simpler and requires less technical knowledge. However, a large element is still missing from our wind material: the primary wind movement; in our case, that will be the plant swaying back and forth in a set direction.

What we are going to do is create a gradient over the Z axis of our asset, which is the up vector in Unreal Engine, and use that as a mask so that when we add a rotation to the asset, the top will move, but the bottom will stay still.

In a lot of cases, vegetation does not get placed down as individual assets but gets either procedurally distributed or painted with the Unreal Engine Vegetation tools; these tools usually instance the foliage, meaning that instead of looking at each actor, it will group up similar plants – for that reason we need to sample the position before any of that happens, which is precisely what the Pre-Skinned Local Position node does, so search for PreSkinnedPosition and put down this node.

As mentioned before, we need to have this node affect the Z axis, which corresponds to the B channel within the material editor (XYZ equals RGB), so we will also need a Mask node isolating the B channel. Plug the Local Position node into the Mask, and to extract a gradient from this, we need to divide this by a value, put down a Divide node, and put the Mask into the A slot. Create a Scalar Parameter and put this in the B slot. I called my Parameter WindSwayHeight and set the default value to 30.

This concludes the basic setup for our gradient; we can add slightly more control by running this through a Power node, which boosts the value you give it in the Base slot. By whatever you put in the Exp slot. In our setup, we can use this to control the gradient contrast. Put down a Power node and a Scalar Parameter called WindSwayContrast with a default value of 2, and do not forget to add all the added parameters to the Wind group so they are all collected in the same place in the Material Instance, as a fail-safe we can run this whole logic trough a Clamp node, ensuring our values never exceed 1 or go below 0 as this will have a negative effect on the math. See Figure 7.4 for the finished graph

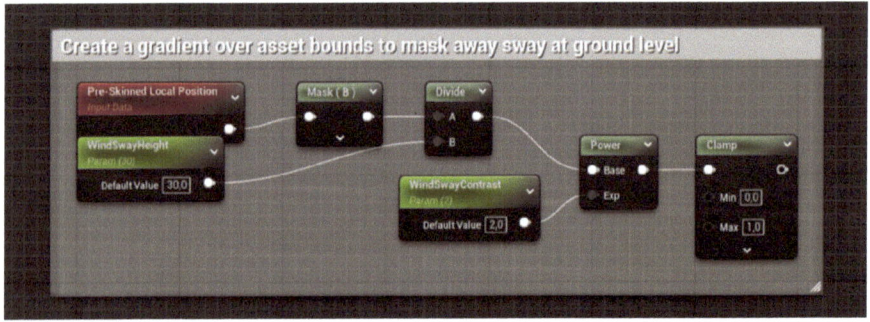

FIGURE 7.4   The graph setup to create a gradient from the bottom to the top of the asset.

Now that we have the mask, we will need to create the logic to sway the asset; we will use Sine waves for this as they lend themselves very well to anything that is supposed to go back and forth. The graph will be a bit bigger than what we have done before, so I do advise to have a look at Figure 7.5 and cross-reference it to make sure everything gets connected the right way. Additionally, you can find the finished material   here: *..\CRC\CreatingHighQualityVegetation\workfiles\ue\CreatingVegetation\ Content\CreatingVegetation\Materials\M_VegetationMaster.uasset*

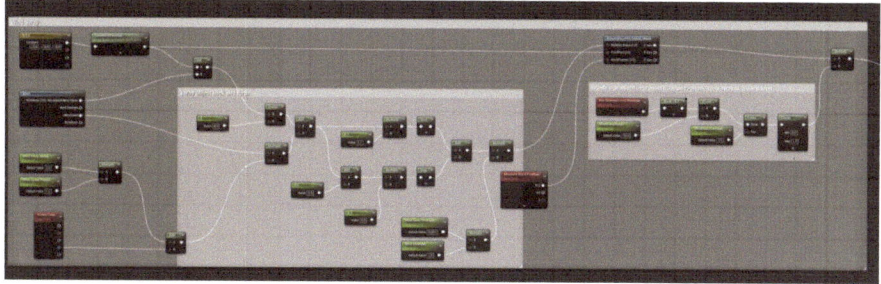

FIGURE 7.5   The graph setup in its entirety.

The RotateAboutWorldAxis_cheap node will run most of this effect, so look for that in the search bar and put one down. Multiply this node with the previously created gradient, and then let's have a quick look at the inputs required.

Rotation Amount (S)

This decides how much the asset will rotate. This is the value we multiply with our gradient, meaning that this effect will be stronger at the top of the asset.

Pivotpoint (V3)

It takes in a Vector3 representing the pivot point the object will rotate around.

We will want the pivot to be the origin of the asset, or 0,0,0 – this is, however, local to the asset, therefore Local Space, and we will need to convert this to the AbsoluteWorldSpace to make sure it grabs the pivot at the location where the asset is placed. We will put down a Vector3 with values of 0,0,0 and run this through a TransformPosition Node, set the source to Local Space and the Destination to Absolute World Space, then plug this into the PivotPoint input.

WorldPosition (V3)

This requires a position to be put in; luckily, Unreal Engine has just the node to do that.

Put down an AbsoluteWorldPosition node and plug this into the WorldPosition input. That will cover one of the three inputs, shaders have never been this easy.

This leaves us with just the RotationAmount left.

We will base our RotationAmount on the built-in Wind feature of Unreal engine; that way, we can, if we wish, more easily integrate it into other Unreal Engine systems later. We can access this by putting down a Wind node. You will notice this has four outputs: the Normalized Wind Vector, Wind strength, Speed, and Actor – we will only be using the Vector and Speed; we are not using a specific WindActor, and the strength is controlled by our own parameters.

For our wind to affect our assets correctly, we need to get a Dot product of our PivotPoint and the Wind Vector; the DotProduct expression computes the dot product, which can be described as the length of one vector projected onto the other or as the cosine between the two vectors multiplied by their magnitudes. Many techniques use this calculation for computing falloff. DotProduct requires both vector inputs to have the same number of channels. I always have to refer to the Unreal Documentation to somewhat understand it so don't be discouraged if you don't, all you need to know is that we need one.

On that subject, if you ever need more information on a node, you can hold down Alt and Ctrl and hover over the node. This will give you more info and also allows you to open the official documentation; this is a great way to see use cases for certain nodes and I do recommend spending some time reading it.

For now however, take out the TransformPosition that was plugged into the PivotPoint of the RotateAboutWorldAxis and also plug this into the A channel of a Dot node, take the Normalized Wind Vector and plug this into the B node, then Divide the Dot Product by 4 – this will create a bit of a falloff effect for our wind, so it does not stop too harshly when reaching the end of a sway.

We will need to add our blue vertex color offset to this logic to ensure that all clusters behave independently from each other. So, put down an add node and plug the divided Dot product into the A slot.

We will plug some more logic into the B slot, so let's create that logic first:

Put down 2 Scalar Parameters and call one "Wind Sway Speed" with a value of 2. This will give the swaying effect its speed and Multiply this by another Scalar Parameter called "Global Wind Speed" with a default value of 1; this does not do anything right now, but it allows us to edit the wind speed for multiple plants that use the same parameter, this is again a precaution in case we want to control the wind at some point globally.

Plug the multiplied values in an Add node's A slot and put down a Vertex Color Node and plug blue channel into the B of the Add node.

Put down one last multiply node, plug the WindSpeed of the WindNode into the A slot, and multiply this by the Vertex Color logic we just created – this will ensure the Wind is affected by our vertex colors, and put this multiply into the B slot off the add node we created in the previous step. For reference to what we just created, look at Figure 7.6.

Now all that is left is the Sine logic to take care of the actual swaying effect, in order to create this in a more natural-looking way, we will create 2 Sine waves and add these together for the final result.

Take our latest Add node and plug it into the A slot of a Multiply node. Then, multiply this by 0.1 and plug it into a Sine Node. This node has many use cases as it outputs a range from minus 1 to 1 and repeats this, which is very well represented by an object swaying back and forth.

This concludes our first Sine logic, so plug this into an Add node, and let's look at the second. Instead of plugging our original Add node into a multiply, this time, we will plug it into an Add node, adding a value of 0.5. This offsets the initial value by 0.5 so

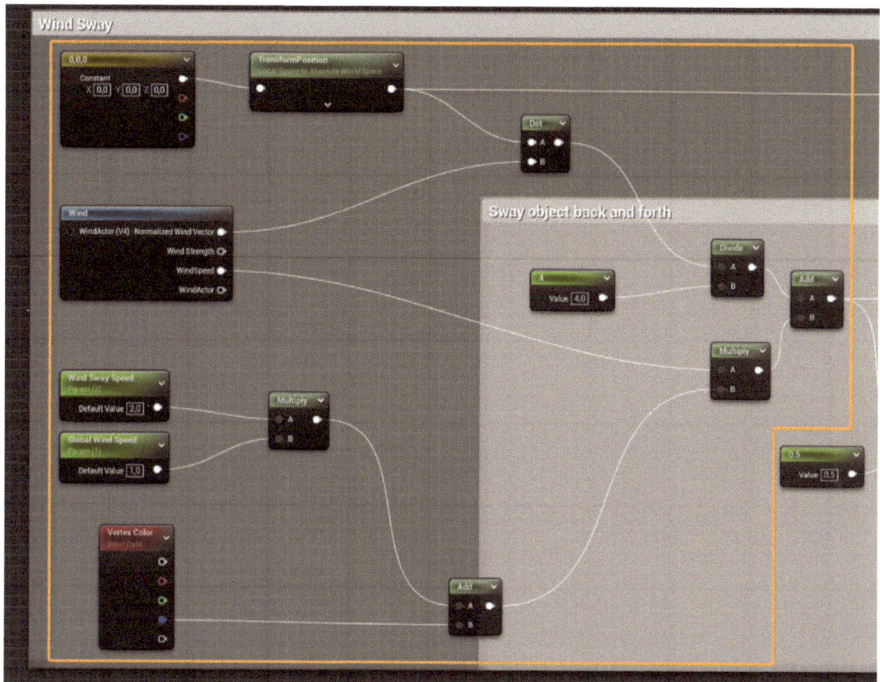

FIGURE 7.6 Graph network in Unreal Engine showing the creation of a sway offset and vertex colors affecting said logic.

we do not create two identical sine waves. Multiply this by 0.3, and then put down our second Sine node.

With both our Sine waves created, we need to add these together, so put down another Add node and plug our first sine in the A slot and second in the B slot. Then, plug the whole chain into a Multiply node. What we also need to add is a global control similar to the Wind Sway Speed, except this time for our strength.

So once again, put down two scalar parameters, called the first Wind Sway Strength with a value of 0.001 and the other Global Wind Strength with a value of 1; this is also connected to our secondary wind strength, so we can control both of these at once in the material while maintaining control over both individually. Multiply this together and multiply the combined parameters with our wind sway logic. Take the final multiply and plug this into the RotationAmount (S).

This concludes our primary wind. All that is left is to combine this with our secondary wind created at the start of the chapter. So go ahead and put down an Add node. Plug the primary wind logic into the A slot and the secondary wind into the B slot. Save the material and look at your asset. You should now see both the leaves wiggle in the wind and the whole plant swaying back and forth. You can refer to Figure 7.7 for reference.

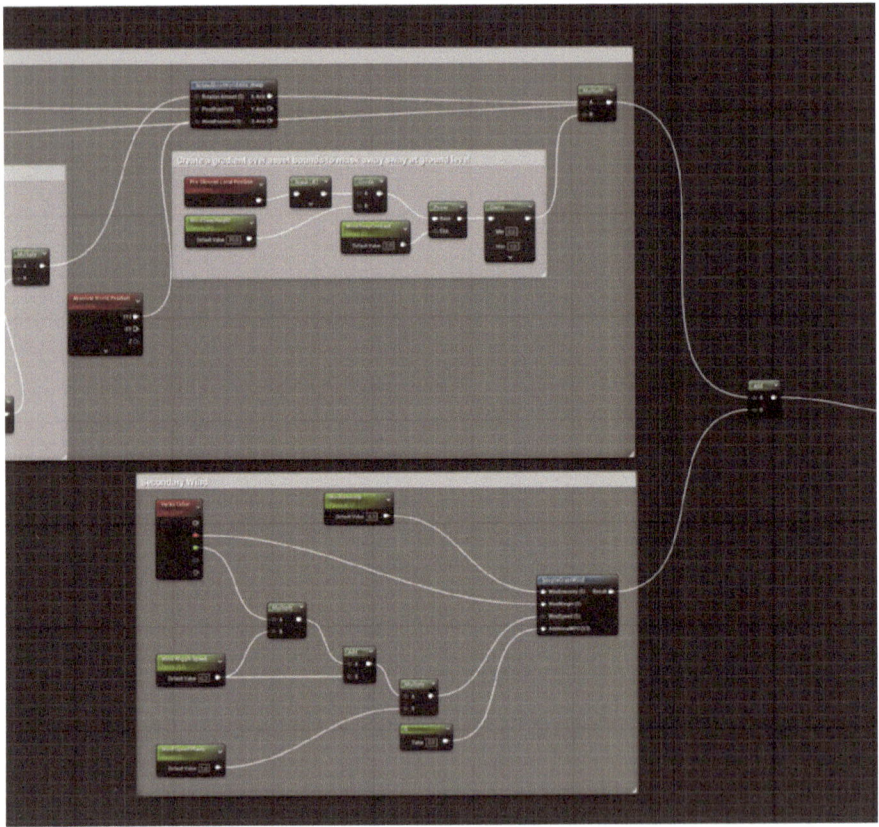

FIGURE 7.7   Node graph showcasing combining the primary and secondary wind.

CLEANING UP AND ADDING SWITCHES TO THE MATERIAL

With our wind material completed, there is a couple things we can do to make it perform better, and make our material graph cleaner and more manageable, we can add switches to the material logic to turn off certain features when we end up not using them. This step is optional.

Adding a Switch to Our Secondary Wind

To add a switch to our secondary wind, create a StaticSwitchParameter and call this "SecondaryWind" plug the SimpleGrassWind output into True and a Constant1 with a value of 0 in false, set the default value to True.

Previously, we used the Add node to add the primary wind to the secondary wind, which will need to be moved to happen behind the switch now, so plug the switch into the A slot of that add, and create another StaticSwitchParameter, this time call it "Primary Wind Sway" plug the Add node into the True value and the Secondary Wind Switch into false, setting it up in a chain like this will mean we can turn of the primary wind without affecting the secondary wind.

If, however, you want control over all the wind at once, we can create one more StaticSwitchParameter, call this one "Wind," and plug the Primary Wing Sway switch into the True value and a value of 0 into the False channel, again setting the default parameter to True.

Figure 7.8 shows the result of all our switches combined. In this figure, 1 refers to the secondary wind, and the line marked as 2. connects to the primary wind.

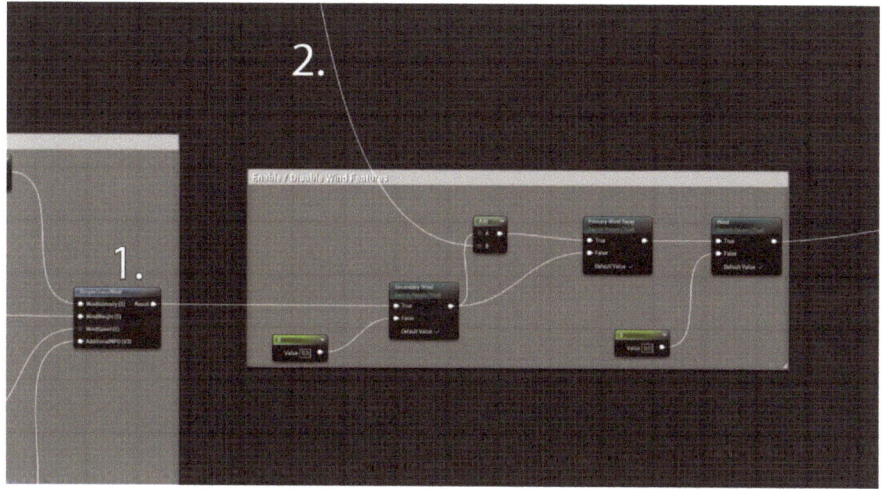

FIGURE 7.8 A node graph showing the switch setup.

In many instances, especially when working on larger projects, logic like this gets reused across all materials in that project. To make that manageable, Unreal has the option to create Material Functions, which allow graph logic to be reused across multiple materials. The benefit is that if you adjust a value in the function, this will get inherited across all the places this function is used. So it is not a bad idea to get familiar with this.

For starters we can create another folder in our Materials folder and call this MaterialFunctions.

In this folder, you can Right-Click and go to Material, then select Material Function. Selecting this will create a Material Function I have called mine MF_Wind. Look at Figure 7.9 for reference, or you can find a version of this in the source data here: *..\CRC\CreatingHighQualityVegetation\workfiles\ue\CreatingVegetation\Content\ CreatingVegetation\Materials\MaterialFunctions\MF_Wind.uasset*

FIGURE 7.9   The content browser inside the MaterialFunctions folder showcasing where to find the material functions.

Double-click the newly created Material Function, and you should see an Output node. The only thing we need to do here is copy in our Wind Logic, so go ahead and open up the M_VegetationMaster, select the first node that is plugged into the World Position Offset, in my case, it is the previously created Wind switch, Right-Click this and press Select Upstream Nodes, this will select all the nodes connected to the selected node, copy this and paste it into the MF_Wind Material Function, plug the Switch into the Output node and hit save.

You can then remove the exposed wind logic from the material and replace it with our Material Function by dragging in the MF_Wind, and plug this into the World Position Offset instead.

The result should now look like Figure 7.10, making our Master Material look nice and neat.

This concludes this chapter and the wind material. Remember that some of the default values used are specific to the plants I tested this on. Feel free to play around with the values to get a result that makes more sense for your particular scenario, you can either do this in the master material or change the exposed parameters in a Material Instance.

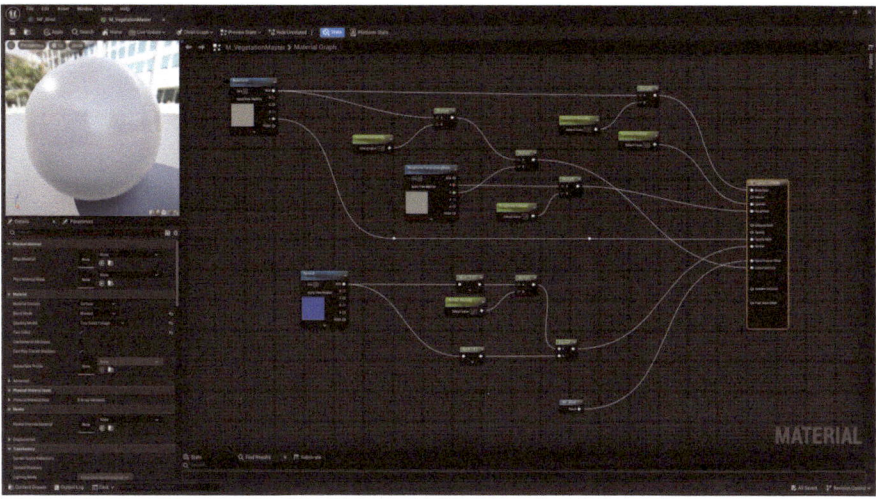

FIGURE 7.10   A node graph showing the complete wind shader with the wind logic tucked away in a MaterialFunction.

Building a Scene

8

With all our plants and trees created, building a scene and showcasing them in context is always a good idea. This makes for a more wholesome portfolio piece and gives you a good idea of how the assets work in context.

In this chapter, I will use all four example plants, two of which have been built throughout the book, and the others are available in the demo files here: ..*CreatingHighQualityVegetation\ workfiles\blender\export* as well as our trees and their variations. For the supporting material, I will use Quixel Megascans assets, available for free on the Megascans website or through Unreal Engine if you have the Bridge plugin installed.

The chapter will explain things to consider when building a scene and how to make things look appealing and natural, but it will assume you get creative to create something truly yours. So, please do not shy away from doing things differently, and strive to improve on what you see in the book rather than copy it one by one.

ACTIVATING THE QUIXEL BRIDGE

As mentioned in the introduction, we will use some external assets to help complete our scene. To do this, we will use Unreal's Bridge plugin. This will allow us to directly use Quixel assets in our scene without setting up any shaders.

To do so, go to Edit, Plugins, search for "Bridge" in the plugin window, and then activate the plugin. Refer to Figure 8.1.

At the time of writing this book, the Quixel Bridge is being replaced by the FAB marketplace, if you are reading this book while that migration has already happened, please refer to the FAB marketplace instead of the Quixel Bridge.

BUILDING A TERRAIN

One of the most important aspects of any scene is the terrain. As with almost anything related to art, it is better to have something on paper than nothing, and building a quick terrain is an excellent way to do this. To do this, we can start by creating a new Level in Unreal by going to File, New Level.

DOI: 10.1201/9781003492283-8

FIGURE 8.1   The plugin window.

I called my level "Main," but you can name it however you want. I selected the basic level template, which gives me a default skybox suitable for our purposes. This Level template also has a default floor object. You can remove that, as we want to create our own terrain.

WORKING WITH A LANDSCAPE

In the top left, you can switch Unreal from Selection Mode to Landscape Mode, or use the Shift+2 Shortcut. This will give you a preview grid of the landscape that will be created. Refer to Figure 8.2.

There are some settings here that have a significant impact on how the terrain is generated. We can leave most of the settings default, but if you are interested in the technical aspects, there is more documentation available here: https://docs.unrealengine. com/4.27/en-US/BuildingWorlds/Landscape/TechnicalGuide/.

The only setting I like to change is the Number of Components. By default, this is set to 8 by 8; I have set mine to 16 by 16.

This gives us a much larger terrain. It will be wider than the area we intend to focus on, but it gives us room to play around with. If you are doing this for a project that needs to perform well, this is an unwise decision, as a larger terrain impacts performance. However, I am not concerned about the terrain's performance for this project since the goal is not to create a game world but to focus on a couple of beauty shots.

With the terrain selected, in the Details panel, you will notice it is asking for a Landscape Material; these materials can be very complex, allowing you to paint multiple

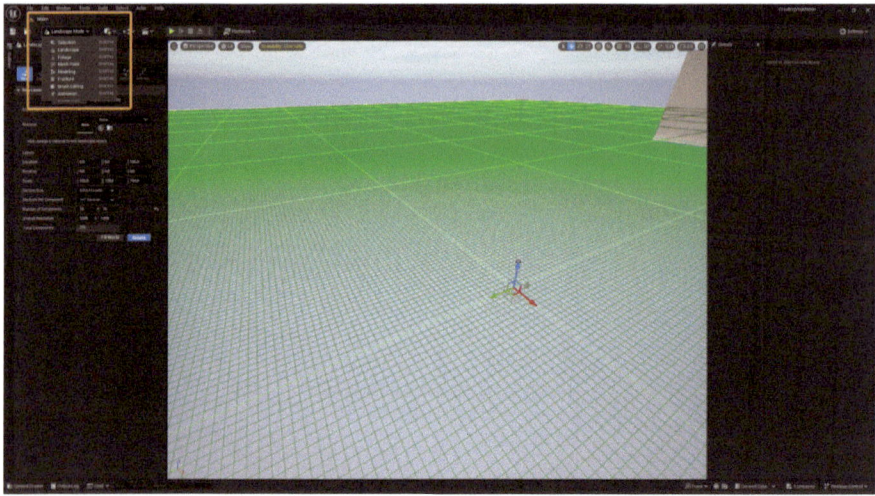

FIGURE 8.2 The landscape mode in Unreal.

textures on the terrain or automatically texture it based on elevation or slope angle. But for our purposes, we are not required to go to this complex, so for now, we can fetch a material from the Quixel Bridge and use that. Refer to Figure 8.3.

1. In the top left corner, where you usually add new content like lights, shapes, or cameras, after enabling the plugin, you will now find the Quixel Bridge. When you open that, you can log in using your Epic Games credentials; alternatively,

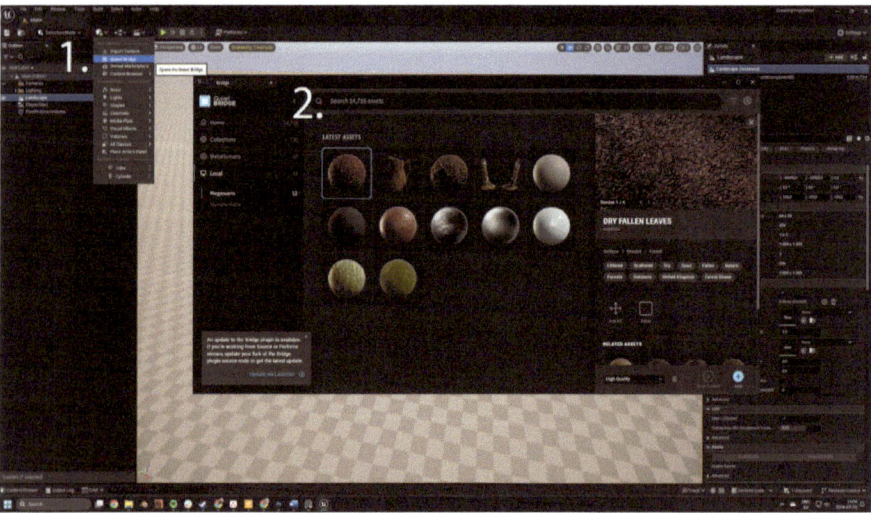

FIGURE 8.3 The Quixel Bridge in Unreal.

you can use the Quixel website, again, at the time of writing his book, it was announced that the bridge is going to be replaced by "Fab" if that has happened when you are reading this, please use Fab instead of the Quixel Bridge.

2. You can browse all collections on the left of this window, but if you want to use the same material as me, you can search for Dry Fallen Leaves and then click Add on the bottom right.

When you press add, it will add the material to a Megascans folder in your project. This material has many parameters that you can change to adjust it to your project's needs. You can access these settings by double-clicking the Material Instance seen in Figure 8.4. I have adjusted the tiling values to 0.5.

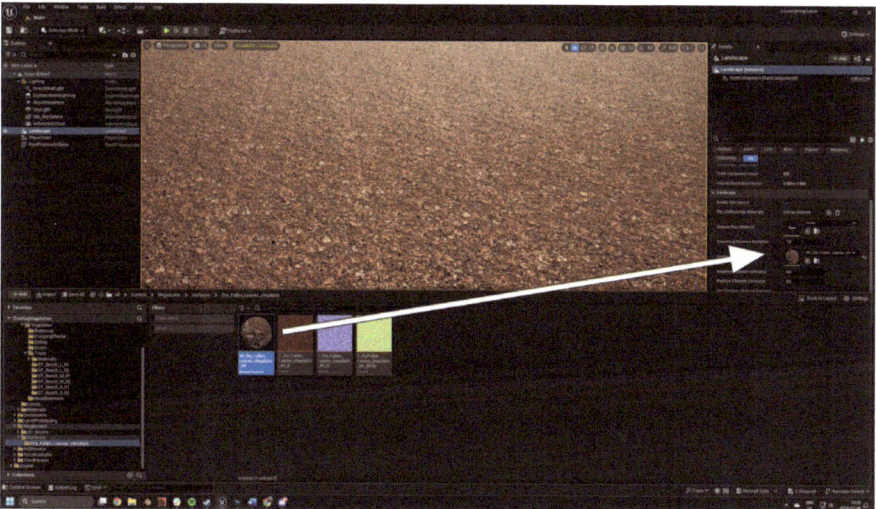

FIGURE 8.4   Unreal editor with the content browser and landscape material slot visible.

Additionally, you cannot just drag the material onto the landscape in the viewport but must pull it on top of the material slot. So open the Content Browser using the Ctrl+Space bar and navigate to the Megascans folder. Find your material in the Surfaces folder and drag it on top of the Landscape Material slot. This should make your landscape appear textured.

If you switch back to Landscape mode using Shift+2, you will also notice you have a bunch of sculpt brushes. If you have never seen these before, I recommend clicking them all and trying them out on your landscape. The best way to get a feel for them is by experimenting.

They essentially allow you to edit the landscape using different types of brushes. However, due to the nature of how landscapes work (heightfields), it is only possible to sculpt the mesh up and down.

At this step in the scene-building process, I keep my brush sizes very large and try to get some general shapes going. I have chosen to make a hill in the middle and build my scene on that. That way, when we place trees further away from the center, they will be placed on lower terrain, bringing more of the canopy into the line of sight—which will help block the view in the distance.

You can refer to Figure 8.5.

FIGURE 8.5   The landscape editor with a hill painted in the middle of the level.

1. In this detail panel you will find all your landscape settings, I have used Sculpt and Noise to get the results visible.
2. The hill in the middle of the map.

As with everything in this book, I would not try to replicate these things one-on-one; if you would rather have a valley or a smaller or bigger hill, I encourage you to play around with that.

With our hill created, or once you are comfortable with the tools, we can move on to the Foliage Painter.

FOLIAGE PAINTER

The Foliage Painter is a tool for painting vegetation on top of terrain and meshes; you can open it in the top left or use Shift+3.

Due to the nature of the painting tool, you might have to remove and repaint certain sections when you make changes to the rule set, so I recommend not getting too attached to the painter's results initially. We will use this to set up some broad strokes, but we will fine-tune it with hand-placed assets later.

The first thing we need to do is drag some assets into the foliage editor. This will allow us to set up the rules. If you are unfamiliar with the tool, I recommend starting with just one asset and playing around with the rules until they make a little more sense. It is a common theme in this book that I make the suggestion to try things on a smaller scale, the reason for that is so you can keep the time it takes to learn and iterate very short.

However, for simplicity and speed, it is best to drag in multiple asset types at once, so if you have your variations made or are using the ones provided in the source files, go ahead and drag each of them into the Foliage Painter.

To do so, open the content browser and drag the largest Beech into it. I like to start with the largest asset first and work my way down to smaller assets. Refer to Figure 8.6.

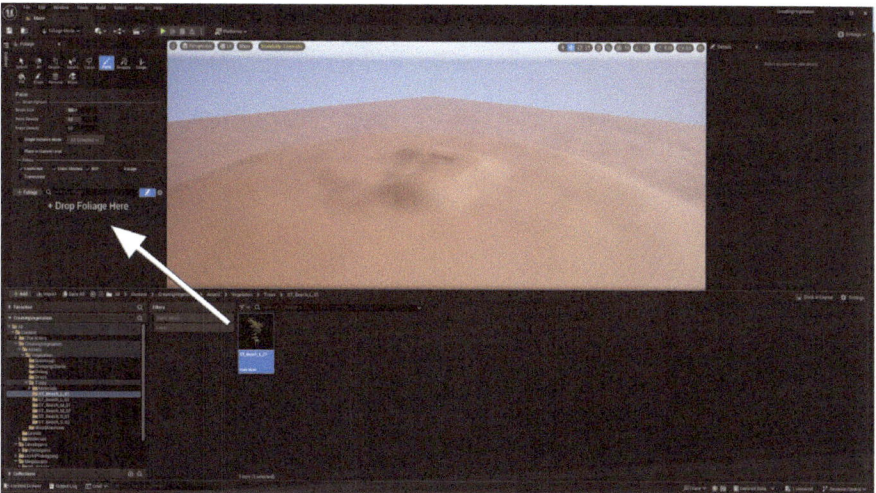

FIGURE 8.6   An example of how to drag an asset into the foliage editor.

Once you have dragged your asset in there, Unreal will make it into a FoliageType, a FoliageType contains settings that the foliage painter can use and consequently these can be set from within the Foliage Painter.

The first thing we can do is increase our Brush Size. You can do so at the top of the Foliage Painting Tool; by default, it is set to 800, but for the size of our landscape, I recommend something like 2500. If you start painting a little in the viewport, you will notice it is painting too many trees. To change this, we need to go to the Painting Tab and change the density. To do this for all trees, click the first tree and then Shift+Click the last tree; they should all be outlined.

When you change values now, it will affect all the trees; I have set the Density to 1.

After every setting, it is best to paint some on the terrain to see what has changed, then undo this with Ctrl+Z or by painting over the existing assets while holding the Shift key.

It is always nice to have some random variation in sizes. With the scaling set to Uniform, you can change the Scale X values to achieve just that. Remember that being too aggressive with these settings can result in issues where assets have a large-scale difference among themselves. So, as a rule, I only scale assets between ten and twenty percent. To achieve this, you can set the Min to 0.9 and the Max to 1.1, which means that the largest and smallest versions can only be twenty percent apart.

You might also notice that the trees follow the angle of the terrain. This is unwanted and can be fixed by turning off Align to Normal. However, this does make the whole thing feel very straight, so under Random Pitch Angle, you can set a low value like 3 to ensure that not all trees grow in the same direction. Refer to Figure 8.7 to see the settings changed in the Foliage Tool.

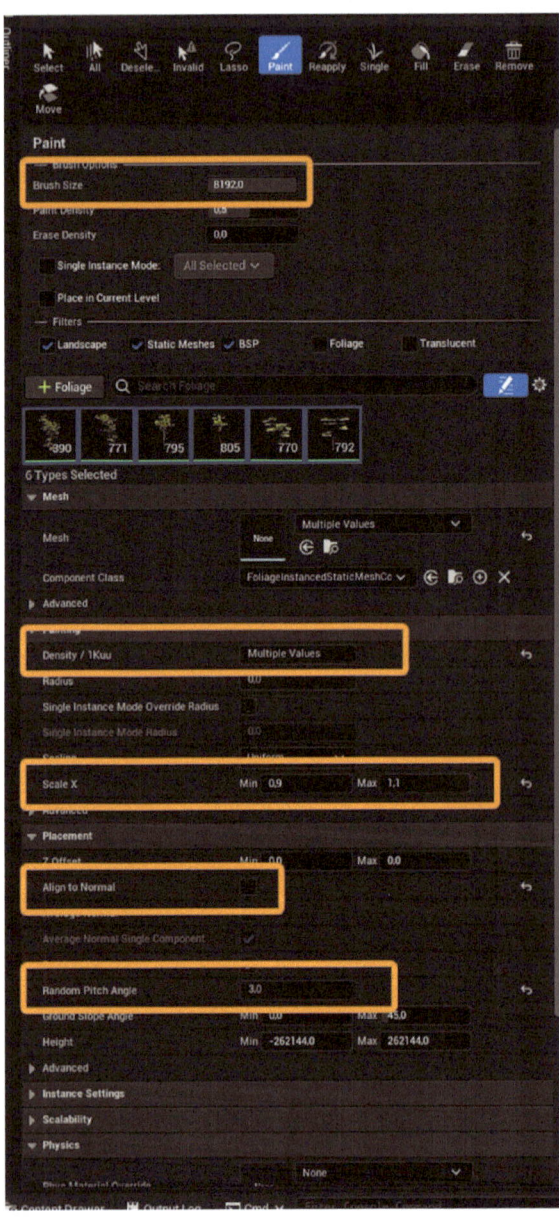

FIGURE 8.7 The settings in the foliage painter tool.

That concludes the bulk of the painting setup for the trees, and you can now fine-tune a bit on a per-tree basis. I, for example, lowered the density of the largest tree to 0.5 to make it feel a bit rarer and more majestic when it does show up, and I increased the density of the sapling to get some more growth happening in front of the player camera. This is an iterative process and can take up to a couple of hours before you are happy with it.

SETTING UP CAMERAS

We currently have a very large landscape, and it is not impossible to populate this all with the foliage painter tool. Still, it does require much work, so more often than not, it is best to focus on smaller areas and make these look very good, especially if you are doing it for just a few portfolio shots.

The best way to do so is to place cameras in your Unreal scene to indicate areas of focus. To do so, select the add button to the top left of the viewport, but instead of Selecting Bridge, select Cinematic and then Cine Camera Actor.

With the Camera Actor in the scene, you can pilot this camera and place it in a position that you see fit. Click on Perspective, and under Placed Cameras, you can select the newly created camera and fly around the scene with it. Refer to Figure 8.8.

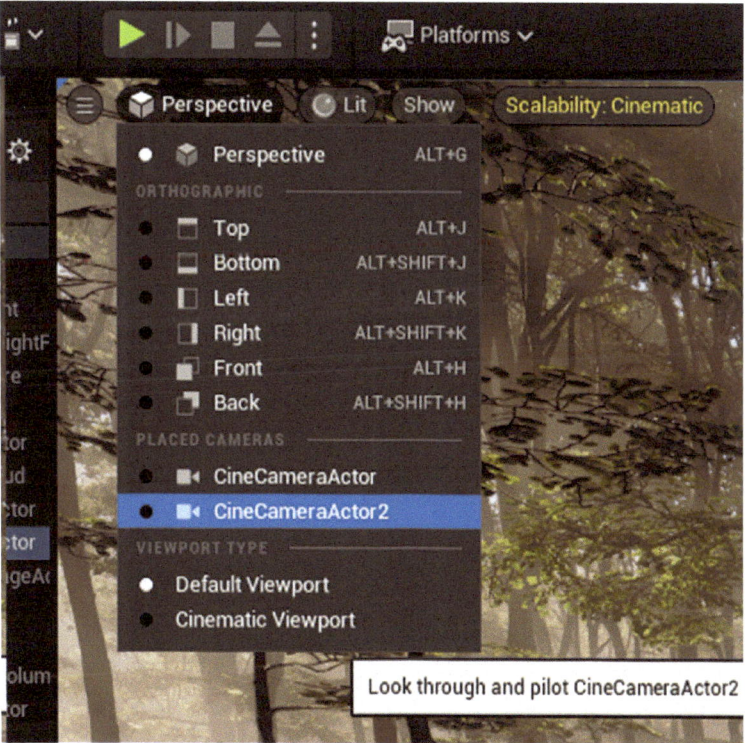

FIGURE 8.8 Example of where to pilot a camera.

The camera's focal length is 35 by default, which makes it feel a bit zoomed in. This does add to the cinematic effect but is not wanted for this camera, so select the camera and, in the Details Panel, change the Current Focal Length to a wider angle. A value of 18 worked well in my case, but you are welcome to use different values.

So, how many cameras should you place, and what should they focus on?

It will be very hard to copy the scenes and cameras showcased in the book, as many of the factors are procedurally driven and depend on the asset, terrain, and personal preference. But that also makes this a very exciting step, as you will be able to explore your own created world and find areas of interest that you would like to flesh out further.

Right now, we have just a terrain and our trees painted on top of them, but you will find that if you squint your eyes, it is already starting to look like a forest. One of my favorite things to do is press play and walk around the forest, use the pilot feature to fly around with a camera to find interesting areas, rotate the lights, or change the intensity, which we will explore in further detail later in this chapter, basically find locations that peak your inspiration and that you feel good about, place down around three cameras that you intend to use as your main cameras, these shots should focus on specific things, a good overview of all your assets combined, a close up on the ground perhaps which showcases all the hard work that went into the ground cover, or maybe a shot from above showcasing both the tree canopy and the ground cover, the amount is limitless and is dependent on you, but I do recommend finding a focus and sticking to it.

In the book, we will focus on one camera, which will showcase all the assets created and techniques learned. However, I encourage you to expand on this and use your creativity to find exciting camera locations and flesh them out. Refer to Figure 8.9 to see what that looked like when I found the location in my scene.

FIGURE 8.9   beauty shot of what the main shot initially looked like.

Lastly, it is important to lock the camera in place once you are done piloting. To do so, Right-Click it in the Outliner, go to Transform, and say "Lock Actor Movement."

Since everything we do will be based on these camera angles, it is important they remain in place and all edits are intentional. It happened to me many times that I forgot I was piloting the camera, flew it somewhere else, and ruined my original shot, so I highly recommend locking them when you are not working with them.

With the camera locked, navigating your scene and selecting things is harder. To remedy this issue, I always open a second viewport; you can do this by going to Window in the toolbar, selecting Viewport, and then selecting Viewport 2. When doing this, remember that Unreal now has to render everything twice so that it can impact editor performance. To minimize stress on your GPU, you can temporarily scale the second viewport down, have one of the viewports in an unlit mode, or temporarily decrease your Scalability Settings. For most of the editor work, I have set mine to High, and I switch to Cinematic when I take my high-resolution screenshots,

GLOBAL LIGHTING

In Figure 8.10, you might notice that some work has already been done on the lighting, so let's have a look at that. Since we have loaded in a default map template, you should have a DirectionalLight, a SkyLight, and an ExponentialHeightFog. The lighting achieved in the images has only used these three elements.

The SkyLight is the light that takes care of all the ambient light, and its color largely defines the color of your shadows. I avoid having my shadows be a grayscale value as this

FIGURE 8.10 Example of where to lock the camera actor.

rarely happens in real life, so I opted for a slightly colder blue hue; I would be very subtle when doing this as it is not something that you want to become the main thing in your scene.

Additionally, I lowered the Intensity Scale to 0.6 to achieve darker shadows and a larger contrast in my lighting.

The DirectionalLight is best compared to the sunlight in your scene, as that is what it imitates. You can actually make it imitate the sun by setting the Lux value to a hundred thousand (which is the same value as the sun on a sunny day), and Unreal will automatically pick this up and adjust your post-process settings. This is useful if you are doing a highly realistic scene and want to make sure you are using physically accurate values; this is slightly more advanced, so feel free to keep yours around the default values.

The angle of the DirectionalLight is also one of the major elements that decide how your scene will look, and sometimes, the best way to find your angle is to play around with many different ones. You can either manipulate the rotation of the DirectionalLight directly, or anywhere in your level press Ctrl+L then keep holding down Ctrl, this will link the DirectionalLight to your mouse and moving your mouse forwards and backward will make the light angle rotate vertically and left or right movements with the mouse will affect the horizontal rotation. I prefer this method because it allows me to select a location and quickly rotate the light to see if I am getting any results I like, and from there, I can refine its position.

Lastly, the ExponentialHeightFog: this asset contains all settings regarding fog in your scene, fog can make or break a scene so use it with caution, you see it a lot in movies and therefore it might appear good looking when overdone, but I find that fog is something usually better done in moderation, it is a great tool in composition as well as it allows you to separate the fore and background very easily, you can see a good example of this in Figure 8.11.

FIGURE 8.11 Left side of the image fog is off and on the right, it is on.

A way to make this fog look more interesting is to add crepuscular rays, better known as "God Rays." This works especially well in forest scenes as the canopy does a good job naturally breaking this up, resulting in some interesting shapes happening across the scene.

For Unreal to create those rays we need to work with Volumetric Fog, which is an accurate way of representing fog from all angles. With the ExponentialHeightFog selected, you will find Volumetric Fog settings in the details panel, start by turning this on.

This effect depends on several factors, including the light angle and fog density. The fog density can also be changed in the ExponentialHeigthFog. I have set mine to 0.08 for this image, which is slightly exaggerated for demonstration purposes. You can play around with this setting until you find a value you like.

The other setting that heavily impacts this can be found in the DirectionalLight and is called Volumetric Scattering Intensity, I have set mine to 8, but for this value, the same principle applies; it works well in my case but it might not in yours so consult your references or find some and settle on a value that works for you, you do not need to nail the perfect value straight away, in fact, I recommend changing and fine-tuning it as you go alone.

Sometimes, it helps to set both these values very aggressively so you can clearly see what is happening and then lower the values once you are happy with the direction and overall look.

COMPOSITION

When it comes to building scenes and scene composition, there are many resources on how to make this visually pleasing, and in my experience, this is something an artist never really stops developing. It is not something you will learn by following along with this book, or any book for that matter. It is something that matures together with your artistic ability and needs to be repeated many times for anyone to become good at it.

However, the important thing, and it should be one of the main takeaways from this chapter, is that good composition comes from doing things with purpose. Everything on the screen should serve some sort of purpose. When using procedural tools, this becomes especially important as they do a lot of work automatically. Often, that results in the purpose being neglected, resulting in mediocre-looking images. The tools are great for what they are, but we cannot let them drive our artistic choices. Composition is a recipe and should be repeatable.

It is about creating structure and following rules as much as it is about breaking them, a dance between psychology and artistry, about introducing chaos in an otherwise structured image, about tickling the viewers' brain, making them look again or, in some cases, make them look away. The best images are images that make you discover something new every time you look at them; in my eyes, the composition is one of the most fun and interesting things to do, and while you are at it, I recommend you to be bold and brave, allow yourself to discover and explore different ideas.

For the reasons explained above, I will not focus too much on a single composition in this chapter but rather explain some rules that I tend to follow and why. You can then use these rules, expand or subtract from them, and apply them to your own composition.

GROUNDING ASSETS

One of the easiest rules to follow, and something that applies to pretty much anything, not just vegetation, is that things look a lot better when they are grounded; assets do not like to be lonely.

I also find that if I start by grounding my assets, the rest easily follows along with that; it usually sparks inspiration and allows me to iterate on what I am seeing. To show you an example, have a look at Figure 8.12.

FIGURE 8.12   Composition example with just the trees and some sticks scattered around.

In this example, you will see a couple of trees scattered around and a whole bunch of sticks. The image does not contain many visually pleasing elements. The trees show a harsh intersection with the ground, commonly perceived as an error. Using grounding elements will help make this image more visually pleasing and fix that error.

I chose this camera angle because there is a nice centerpiece with some interesting shadow play. The trees are already somewhat spread out across a mid-ground and background, a concept we will learn about later in this chapter.

If you look at Figure 8.13, in this example, I have added all the Wood Anemone variations close to all the trees. The trees now appear more grounded, and the hard seam where they intersect with the floor is no longer visible; this is a good example of

FIGURE 8.13 Composition example where the wood anemones have been added, fixing some issues and introducing new ones.

a ruleset. We could, for example, say that the wood anemone should always be placed close to trees. They tend to grow across the forest floor in their natural habitat, but we can bend that to create a visually pleasing image.

We notice that this grounds the tree, but the group of trees and anemones now appear to be floating around in the scene. There are three things I would like to do to fix this. The first and most straightforward one is to ground this as well, with a different asset or ruleset. Secondly, we can already start thinking about breaking our own rules and scattering some Wood Anemones in places where they are not immediately expected. Lastly, we can add variety.

As seen in Figure 8.14, when we add some grass to the mix, everything starts looking more organic and natural, the previously isolated look is starting to break a bit, the asset borders become less visible and we get a noisier look, which is somewhat typical for forest floors where a lot of things happen all at once. Having a ruleset in mind when doing these things and applying it consistently allows you to bring structure to that chaos, something explained in further detail earlier in this chapter.

Now is a good time to start applying similar techniques to the rest of the forest floor; in Figure 8.15, I use the Creeping Charlie, you can find it in the example files here if you have not built them yourself ..*CreatingHighQualityVegetation\\workfiles\\ blender\\export* This is a nice low to the ground asset that we can use as an underlying layer, or of breakup of the brown forest floor, In the example you will notice I have painted them in the corner in a somewhat sweeping pattern, this helps hide the corner of the image and functions as a visual container, visual containers will be explained in more detail later.

Additionally, I have used them to blend the other groups together a little, it is easier to make it appear as a cohesive image if the colors are somewhat similar; previously, the

FIGURE 8.14 Composition example, where variety is added to increase visual fidelity.

FIGURE 8.15 Composition example.

green and the brown had a strong contrast, resulting in a floating look, with the Creeping Charlie in the mix the colors are more cohesive and the forest floor starts to look more uniform and natural.

I still left some open spaces for different assets and to create natural paths.

Using the space left empty for the natural paths, in Figure 8.16, you will see that I have filled this up with some scattered dead leaves; these assets are downloaded from the Quixel Bridge, how to do so has been explained earlier, but as a quick recap, you can

FIGURE 8.16 Composition example where leaves are added in the paths left by the Creeping Charlie.

open up the Quixel Bridge from within Unreal Engine, download the assets you would like to use, and quickly add these to your own project. I recommend making most of the assets yourself and using this to fill in the gaps; that way, you ensure it will make for a stronger portfolio piece.

Adding the leaves further contributes to blending the ground together and introducing some depth, shadows, and color variation.

Now with the major elements in place, and our ruleset setup. It's time to break it a little.

In Figure 8.17, you will see an example of this, you will notice that in later stages of creating an image, iterations will become smaller, and in this instance, the only thing I did was add the Wood Anemone in some places outside of the ruleset, sprinkling it here and there makes the image more dynamic and alive.

With the ground set up, there are many things you can do to finalize this image, play around with the lighting, grab some rocks from the Quixel Plugin, add some mushrooms, sculpt the terrain, or add more assets and repeat a similar process as shown here, in Figure 8.18 I have added some of our smaller Beech trees to help tie the image together.

There is still much to be desired for the background of this image, and at this point, I do not consider it done, but it is in a good state to start collecting some feedback from friends or colleagues or move on to work on a new shot to see if you would like that better, I do not like to work on things from start to finish in one sitting, and always try out a couple of things before settling for the final result and I encourage you to do the same, a good number of shots would be around six to eight that way you can select the best ones and polish those up.

With that being said, there are a couple of other things to keep in mind when building these scenes. We briefly touched on them in these examples, but let's take a deeper look at Visual Containers.

FIGURE 8.17   Composition example.

FIGURE 8.18   Composition example.

VISUAL CONTAINERS

The term Visual Containers is somewhat self-explanatory. The intention is that they contain the eyes inside the image. As described before, a composition should serve a purpose, and the most common purpose is showing the viewer something exciting or visually pleasing. However, this is sometimes easier said than done due to human nature and short attention spans.

For that reason, composition is often used to help the viewer see what was intended in the image.

Every image we produce on a screen or canvas is square, so first and foremost, we want to contain the viewer within that square. This is true for paintings, computer screens, and phones; we want viewers to immerse themselves in the image. Noticing that border breaks immersion, and worst-case scenario, makes the viewer notice something off-screen and get distracted. I find it easiest to showcase this in an abstract.

If you refer to Figure 8.19, you will see an empty gray canvas; hopefully, you will notice that it is quite easy to see the borders and notice your eyes drift off the image, this is something we would like to avoid, this is where visual containers come in.

FIGURE 8.19   A gray empty canvas.

Figure 8.20 shows the same empty canvas, but this time, a darker gray border has been added. On top of the border added, two more deliberate choices have been made.

The human brain is somewhat programmed to look at the brightest values first, so I deliberately made the border a darker shade of gray to indicate this is of lesser importance and should not be looked at too much. Secondly, I made the border irregular, straight lines are easier to pick up, look less organic and run the risk of breaking immersion.

What this does is create what I like to refer to as a "True Canvas" and all that is left is fill that with the subjects we want the viewer to look at.

On Figure 8.21, you will notice some white dots. These have a brighter value to draw the viewer's eye but also give clear points for the viewer to look at.

Where to place these points can also heavily influence the image; in the example, they are somewhat scattered, but you can apply the rule of thirds or choose a center composition. This is all about experimenting and figuring out what method works best.

FIGURE 8.20 A gray canvas with irregular borders.

FIGURE 8.21 A gray canvas with irregular borders and focal points added.

It is also okay not to use all the methods, but it is a good idea to keep these principles in mind while working on a composition.

Figure 8.22 shows you a good example of what a visual container can look like in a more refined example.

If you already have some of your shots lined up, now is an excellent time to experiment with visual containers and see what they can do for your image. Remember to play around with different elements; as always, the key is iteration, trial, and error.

FIGURE 8.22   An example of a visual container in our forest scene; the top is without the visual containers, and the bottom is with.

FORE, MID, AND BACKGROUND

Looking at Figure 8.23, you will notice that this image applies all the previously explained principles, there are visual containers present, and some elements are placed around for the viewer to look at. But another thing is very apparent in this image and worth pointing out and exploring a bit further: the fore, mid, and background and how to use them to your advantage.

This is perhaps the most straightforward aspect of composition, yet it is also often forgotten; adding a fore, mid, and background will quickly create some depth. If you intend to have a moving camera, this will introduce an effect known as the parallax effect.

Since vegetation is usually something seen from relatively far (for the size of the subject) it is sometimes hard to display all the hard work you have put into a bark or leaf texture, I find the foreground to be an ideal medium to get some of that shining through still, and I usually add some smaller plants or canopy leaves poking into the frame, showcasing that the vegetation I build is looking good both from a reasonable distance and up-close.

On the mid-ground, I usually have what I consider to be the most important subjects of a particular shot; the beauty shot in Figure 8.23 showcases all the photogrammetry trunks scanned for the book.

FIGURE 8.23   A beauty shot with all principles applied.

For the background, this is a great place to get the viewer wondering about the location. In our case, it is a forest, so I placed a couple more trees, but a distant mountain or desert will quickly change the setting; of course, this needs to be somewhat relevant to the image you are trying to create. I like to keep my background balanced; there should be enough elements to see a thing or two when the eye drifts off, but not so much that the viewer stays there; you could view it as a wall to bounce the eyes off from when they drift off. There are exceptions in case you want to surprise the viewer; for example, you can put a UFO in the background, which would probably be memorable; make sure that that is the purpose then and does not accidentally take away from the main elements of the image.

I find it looks best if the foreground and background are naturally broken up. One good way, as explained in Figure 8.11, is fog, but this is not the only way.

Lighting plays a huge role in any composition and can be a book in its own right. Many online resources explain much more about lighting, but I will keep the explanation here relatively simple to remain focused on vegetation.

What I like to do is something I refer to as painting with light. I do this for almost every shot, and it allows me another layer of control; usually, I work with point lights, and you can try and add one to your shot. It is important to turn off shadows as we do not want to add it as a new shadow caster, but allow us to massage the light a little in our favor. I usually use them to brighten up whatever is in the mid-ground, so it gets a higher value and is therefore perceived as more important. Sometimes, it is also helpful to brighten up overly dark areas in your composition without messing with the primary light sources.

A second lighting technique I use to separate elements is adding rim lights. I highly encourage you to Google this to learn more, but to summarize, this means adding lights to highlight the edges of an object, helping it separate itself from the background. In Figure 8.23, you can see this technique applied to the sapling trunk. Notice the bright white line running across the edge; this is the rim light doing its job.

Something that always stands out is straight shapes. When working with long, straight shapes, breaking these up and creating visual interest is essential. Since we are working with trees, we are constantly dealing with long, straight shapes, and there are multiple ways of breaking them up. Here are some suggestions:

1. The easiest way to do it is to add a shrub or a bush or some other branches crossing over the trunk to create some visual interest.
2. Burls, or scars added as decorations, go a long way in breaking up longer and simpler surfaces.
3. Less obvious would be to add some shadow play to the tree trunk, allowing the contrast of the sun to help break up some of the shapes.

To see some of these principles in action, refer to Figure 8.24.

FIGURE 8.24 Examples of how to break up straight shapes to create visual interest.

1. A relatively boring looking straight trunk, it has little going on except for a texture.
2. Rotating the light so there are some shadows, creating a nice contrast and help break up the straight lines visible, this is in a lot of cases already enough to but you can take this further easily.
3. Adding some branches crossing the shape, introducing horizontal lines as well as vertical lines help take this image from bland, to visually interesting.

Realistically, you will not have time to do a pass on every tree in your forest, and additionally, not every tree in nature grows in a way that is visually pleasing so as always this is a game of balance, stay on the lookout for these things in higher priority areas and try to work in a way that prevents this as much as possible and it should go a long way in making all your images more visually pleasing.

That concludes some of the most used techniques to help guide the eye and introduce visual interest across the scene; if you have been working alongside and have been introducing these elements to your scene, you should have ended up with a couple of images that, in your opinion show promise and could be taken further, if you haven't and just read along, now is an excellent time to spend some time in Unreal, putting everything you just learned to the test.

POLISH

First of all, congratulations for making it this far in the book, if you have reached the polish stages, a lot of work has already been done and you should already be proud of yourself for committing this much time and effort into getting here. With that being said in my experience the polish stages can take up equally much time as everything before it. This is often referred to as the eighty/twenty rule, meaning eighty percent of the work takes as long as the last twenty.

Therefore, in this stage, it is important to prioritize and polish where it matters. This is somewhat personal to the purpose of your project, if you are building a video game you often end up polishing what is considered the golden path, the path that the player will most likely travel on the most or assets that tend to repeat the most, as this will have an impact across the board.

But if you are simply doing this for portfolio work, a good area of focus could be your camera locations, as these will be the shots you intend to upload to your portfolio.

Again, this is highly personal, so it is hard to give solid examples, but here are some examples of what this process looks like for me.

Setting up a Unique Lighting Setup for Each Shot

This is a luxury you have when not doing it for real-time purposes where the main light usually remains in the same location, but if you are doing separate shots, I create a new folder called something along the lines of "Shot_1_Lighting" This way I can toggle the whole folder on and off when switching cameras, but in this folder, I will have a unique

Directional Light, Skylight and any Points Lights used for that specific shot, this setup allows me to be freer with how I art direct my shots and get the most out of each one, and I highly recommend it if you are doing separate shots, remember that you are in a digital environment and the options are almost limitless.

Ask for Feedback

In this step, it could very well be that you have been staring at your images for too long, so it is not a bad idea to step outside for a little, look at some real trees and ask for feedback from friends, peers or family, analyze the feedback and see if you understand why they have said it; then make a decision on how to fix it, it helps if you ask for specific feedback and let people know what state the project is in. Good examples would be "I am finishing off this image, and I would like to see if you see any remaining errors" — that way you let people know that feedback on primary shapes is no longer that relevant, and what you are looking for is errors. In different stages you could, for example, phrase it as "I am trying to get a good camera angle, but feel like it is not really working, what would you do to get this image to look better?" – Again, the goal here is to get whoever gives you feedback into the right state of mind.

I highly advise being specific when asking for feedback, especially if it is in an online community where you don't know people personally. In my experience, simply asking, "Can someone give feedback on this image?" results in vague and unsatisfying answers. Remember, what you put in is what you will get out.

Additionally, never take feedback personally, analyze what is said, thank them for the feedback, and act accordingly.

Fix Your Bugs

It is a widespread term used in the game industry but applics just as much to personal work. While working, you have probably made some mistakes. It is okay to leave those mistakes while you arc working for the sake of moving forward. Still, in the polish stages of your project, it is beneficial to look out for those errors in the images you intend to present and fix or hide them in the image. Hiding them only works if the goal is static images and the issue does not appear across multiple images.

Pay Your Technical Debt

Technical debt is a term less often used but still very common. It refers to everything you shoved forward while thinking, "I will fix this later" or "I will do it like this for now and implement it properly later." Some of this debt will always stay, but some will shine through in the final images. Now is an excellent time to resolve that.

I often forget about bugs and technical debt, so I have made it a habit of writing down anything I notice myself pushing forward for later so that when I get to the polish stage, I can refer to my list and fix anything that is still on there. If you find yourself in a similar position, I recommend doing the same to keep track of it.

If you intend to share your work with others, a polish pass can also include folder structure, naming conventions, and optimization.

Generally, the idea here is to clean up after yourself and wrap things up. I recommend not rushing this step, as it is very important and often makes the difference between a solid portfolio piece and an image that is somewhat falling apart at the seams. Aim to elevate your work from good enough to impressive and well-thought-out.

POST-PROCESSING

We have briefly spoken about the post-process volume before to fix some exposure issues, but I find that starting to work with it toward the end gives more accurate results.

If you have not added a post-process volume to your scene already, you can do so with the Add Content button at the top left of the Unreal Viewport. Under Visual Effects, you can select the Post-Process Volume. This is a volume, so to see the effect you need to be inside of the volume, or you can turn on the Unbound checkbox in the Details Panel, that way the post-process volume will work across your scene.

This is another element that can be very influenced by personal preference, so I recommend you have a look at the settings, play around with them, and see what they do, but do so in moderation. The tricky thing is remaining neutral; once you have set your saturation to 1.5, anything below that will feel desaturated, even though it is not, so slowly massaging the post-process volume until your image has gotten to the desired result is the way to go for me.

Some settings that I enjoy playing around with are the Temperature. I find Unreal Engine standard images are usually slightly too warm, so I set the temperature to something like 7500. This does get you a lot of colder hues, but it enables you to bring some of that back using Point light with warm colors. Still, if you are feeling creative, you can also try it the other way around, making the post-processing a bit warmer and bringing in colder colors through point lights. It is up to you; if you follow along references closely, it would be best to match that.

Secondly, I tend to slightly increase the saturation to bring out more greens or flower colors and increase the contrast. I often use low values like 1.1, as a lot of this can also come from your textures, and I use the post-processing volume only to enhance a couple of things.

You can refer to Figure 8.25 to see an example.

You can consider increasing your highlights to boost the specular response, which usually gives you a more dramatic look, or, under Visual Effects, increasing the sharpening to help bring out many of the details and make the image appear more crisp. You have many options here, and it is best to explore them all to see which ones work best for your images.

To avoid getting lost in all of the settings, I recommend adding them individually and toggling them on and off to see what difference they make and if they are still relevant. Remember that, similar to the lighting, it is okay to use different post-processing on different images, although you have to be careful that the images remain cohesive.

FIGURE 8.25   An example of the post-process volume.

RENDERING OUT IMAGES

Once you are happy with all your images and ready to take these out of Unreal, the best way to do that is to select the three lines in the top left of the viewport and open the High Resolution Screenshot tool, refer to Figure 8.26.

I usually set the screenshot multiplier to a value of 2, but if you have a low resolution monitor, you can opt for a value of 3 or 4. Just keep in mind that this is heavy on your GPU, so make sure to save it before you try this.

FINAL TOUCHES IN PHOTOSHOP

Lastly, I wish someone had told me this when I was starting, but it is okay to do some Photoshopping of your images if you want to enhance specific parts of it, it is of course impossible if it is for a video game, but those images are constantly moving so they require less detail to attention than still images.

Do not be afraid to bring your image into Photoshop and create some adjustment layers for the finishing touches, or use the Spot Healing Brush to remove unwanted errors or lighting artifacts. This is a huge timesaver and allows you to focus on other things.

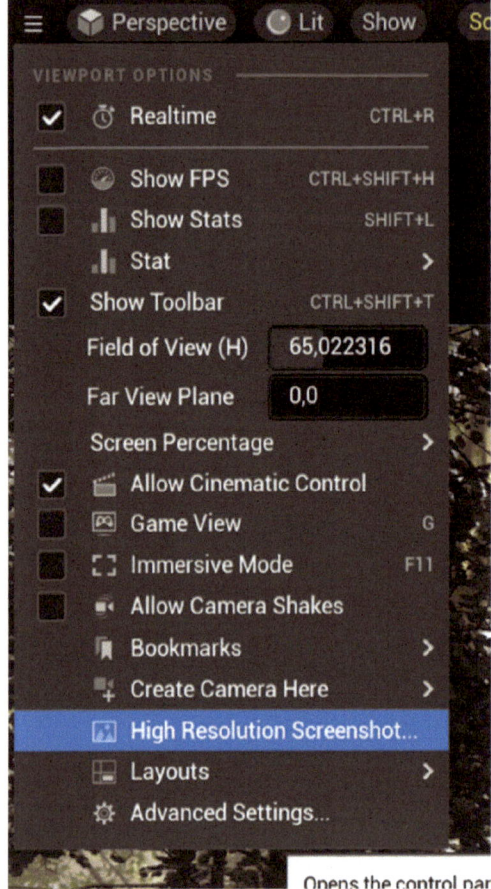

FIGURE 8.26   The viewport settings with high resolution screenshot highlighted.

That said, if you find yourself leaning heavily on Photoshop to make the image look good, you should ask yourself if it would not be better to fix it in the source. But, for the finishing touches, it is perfectly fine.

GOOD PRACTICES FOR YOUR PORTFOLIO AND FINDING A JOB

As this is the book's final chapter, I would like to finish it with general good advice for your portfolio and some example shots to help you along or inspire your own.

If you are early in your career and are looking to break into the industry, you are probably already working on your portfolio. Without one, finding a job will be pretty much impossible, so if you aren't, I strongly recommend you start looking into creating one.

The easiest way to create an online portfolio is to open up an ArtStation account and upload your portfolio work there, which is also an excellent bridge to the first and potentially most crucial point.

PORTFOLIO WORK VERSUS NON-PORTFOLIO WORK

Realistically speaking, as an artist, there is a good chance that you will want to work on many things, and I recommend that you explore every side of it, pick up those ambitious projects, and chase the hype of starting a new piece. However, keep in mind that there is a difference between portfolio work and non-portfolio work.

Portfolio work should be genuinely yours, showcasing the best of your abilities. I always recommend that if the intent is portfolio work, you push the quality to your limits and try to go above and beyond to make it look good. Never rush portfolio work.

Non-portfolio work is, for example, when you have followed a tutorial, learned something new, and ended up with something that is technically a copy of the tutorial you just followed. There are exceptions. Sometimes, you have a limited amount of time; if that is the case, I recommend drawing a line where you say: this is where the learning stops and turn the result of a tutorial (or book) into your own portfolio piece, showcasing that you have understood the assignment and built upon it.

Like a composition, a portfolio piece should have a purpose: to showcase that you are capable and understand how to get to a particular result independently rather than following a recipe.

This has the added benefit of removing some pressure when you are trying to learn. It is often better to deliberately practice rather than try to create a portfolio piece every single time. If you are struggling to make leaves, focus on making 50 leaves rather than building 50 trees; that way, you optimize your learning.

WHAT TO CONSIDER FOR YOUR PORTFOLIO

Who Is Going to See It?

Asking yourself this question is going to help guide and shape your portfolio images; because of the nature of a portfolio, it will most likely end up being other artists, but more importantly, potential employers, so you want these images to look the best they can, but also be as easily findable and accessible as possible.

How Are they Going to See It?

Another important consideration to keep in mind is how they are going to see it. In the best case, they will be on a monitor. Still, there is a chance that whoever sees it will be on the train, in between meetings, or even in the bathroom, perhaps even all three, so it is important to make that first impression last. I always recommend checking your images on multiple monitors to ensure the images looks nice and balanced on all of them, but also check it on, for example, a TV, a TV usually has a lot of image enhancements going on that in a lot of cases do not do your images any favors, make sure to check it in as many places as possible from multiple distances as, but also different lighting conditions, that way you ensure your image looks good no matter where it is viewed from.

How Do You Make It as Easy as Possible for Them to Access Your Portfolio?

For reasons described in the previous paragraph, it is also essential to ensure your portfolio is as easily accessible as possible. When you are using ArtStation, a large portion of this is already taken care of, which is why this is a very popular choice, but there are still some considerations to make on your end if you send out your portfolio link, it is helpful also to make it a Hyperlink, that way they only need to click on it to open up your portfolio, make sure your contact details are readily available and in general show respect for the time of the person viewing your portfolio to leave a better impression.

Watermark Your Work and Name Your Images

This is another thing some people forget, myself included, but it is a good idea to watermark your image with your name and potential contact information. There is no need to make it very large, and it should be the least intrusive it can be. Pick a corner and keep the text small, but in case someone shares just your image or saves the image to their PC, that information is collected as well; additionally if someone decides to share your image somewhere else, you ensure that your name and info is shared along with the image.

Sharing your info that way is also a little safer because search engines cannot read the image, if that is a consideration you would like to make.

Additionally, I recommend naming your image something like YourName_JobTitle_ Subject_01 as image names usually get uploaded and picked up by search engines. This helps when recruiters look for specific things and decide to Google "Vegetation Artist." it will give you a higher chance of being found in those scenarios but also ties into the previous point if someone saves down an image and it is named in a sensible way that gives them information about the author, in some cases a website reprocesses the image and when you re-save it and the name gets lost making it redundant, but this is not the case for ArtStation so still worth your while.

WHAT SHOTS TO INCLUDE
IN YOUR PORTFOLIO

A portfolio is meant to showcase your work in the best way possible, so Beauty shots are a given. However, it is also a good idea to show how you got there and, where possible, show your technical capabilities. This is especially important if you are an aspiring Vegetation Artist; it is both an artistic and technical profession.

So, on top of beauty shots, there are a couple more things I recommend showing on your portfolio:

1. Asset overview shots: these are typically shots of just the assets singled out and do a great job of showcasing the assets used to create the scenes. Additionally, you can make beauty and overview shots of individual assets to showcase them in the best way possible. If you have used any assets from external libraries, this is also a great way to show off what assets you created from scratch and will not leave the viewer questioning what was done by the artist and what was brought in externally.

   You can also include things like your UV layouts or Vertex Colors to showcase that you understand how to build efficient UVs and can set up an asset technically.

   Refer to Figure 8.27 for a good example of an overview shot.

FIGURE 8.27   An example of an overview shot of the Wood Anemone.

2. Vegetation looks best in motion, so I recommend if you have a wind shader setup to create some videos as well, you can do this using the Sequencer in Unreal Engine.
3. Don't forget about the shaders. It is always a good idea to see some shader breakdowns. Just ensure you understand what is going on and that the images explain that clearly. It is recommended that you support that with a short description. I do not recommend making a single screenshot of the whole graph and not explaining anything.

Keep in mind that if you put it on your portfolio, it will most likely be brought up in any job interviews. So, if you, for example, have followed a tutorial and put that on your portfolio, it is important that you actually understand what is going on and are able to explain and/or have a conversation about it.

There is nothing wrong with copying a tutorial to get something to work, but it is not always wise to showcase it on your portfolio, just make those considerations before uploading to your portfolio.

For reference, here are some shots that I consider portfolio ready (Figures 8.28–8.30).

FIGURE 8.28   A portfolio ready shot of beech tree.

FIGURE 8.29   A portfolio ready shot of the Wood Anemone.

FIGURE 8.30 A portfolio ready shot looking up the canopy of the forest.

PROMOTE YOUR WORK

Since most artists post their work on ArtStation, you will be putting your work into a highly competitive environment, so it is often necessary to promote your work a little bit to get some eyes on it. You can use the number of views and/or likes to a certain degree to measure the quality of the work, but you need to have realistic expectations.

I usually share my work on my LinkedIn, and in Discord communities I am part of, this helps to get the ball rolling and hopefully end up on the Trending page on ArtStation, and from there on, the ball rolls itself. You can compare your views and likes to similar work, but I consider it successful if my work gets a thousand views and fifty likes. On paper, this might not sound like too much, but it is very much in line with work from other professionals. There are outliers, but in my opinion, you should aim for realistic numbers.

Also, remember that in most cases, you will need to send your portfolio to the right people. You will not find a job easily by just uploading to ArtStation; it is just a part of the process. If you want it to be seen, send it out.

Therefore, I also recommend supporting yourself by building a professional network. A method that works well is to start early and accumulate this as you go; I have used LinkedIn and tried to add everyone I met at, for example, conferences or online communities. You never know how people will develop and if that can mean something to you in the future; in some cases, you can even contact relevant professionals by sending them a friendly message on LinkedIn. In fact, now would be an excellent time to find me on LinkedIn. Send me an invitation and a link to the work you produced after reading this book. I would love to see it.

CLOSING NOTES

Now that you have reached the end of the book, I want to sincerely thank you for your effort and attention up until this point. I hope you have learned a thing or two and feel more secure in continuing your learning journey or job applications. Do find me on social media and send me any results you might have gotten from following along with the book. I look forward to seeing it.

Enjoy the process, work hard, stay curious, and surround yourself with like-minded individuals.

Wishing you all the best.

Glossary

Add Node: A node used to add two values together in a material graph, commonly used to combine or blend different effects.

Adjustment Layer: A layer in image editing software that applies color and tonal adjustments non-destructively to all layers beneath it, allowing for flexible editing and experimentation.

Albedo: The measure of diffuse reflectivity or reflecting power of a surface.

Albedo Map: A texture map that defines the base color of a 3D model without any lighting or shadow information, representing the true color of the object's surface.

Align to Normal: A setting in Unreal Engine's Foliage Painter that aligns assets, like trees, to the normal of the terrain, making them follow the terrain's angles.

Alpha Channels: A channel in an image file that defines transparency levels, used in textures to determine which parts of the surface are visible or hidden.

Alpha Clip: A mode in Blender and other 3D applications that uses an alpha map to make parts of a surface invisible, based on a cutoff value for transparency.

Alpha Merge Node: A node in 3D software used for combining multiple images or textures into a single output. It merges the Albedo (color) and Opacity maps into one texture.

Ambient Occlusion Map: A texture map that simulates the soft shadows created in crevices, corners, and areas where objects are close together, used to enhance the realism of 3D models by adding depth.

Apical Part: The top part of a plant or tree that typically grows upward.

Automatic Decimation: A process that reduces the number of polygons in a mesh automatically, often resulting in less clean and messier topology compared to manual reduction methods.

Average Normals: A setting in 3D modeling that adjusts the normals (perpendicular vectors to surfaces) to create a smoother appearance by averaging the angles of adjacent surfaces.

Background (Back): The distant part of an image, often used to set the location or add depth.

Bake: The process of calculating and applying textures or lighting effects to a 3D model to pre-render details and improve performance in a real-time environment.

Bake Stitch: A method in SpeedTree for blending different textures and geometries on the tree model.

Baker: A tool in 3D software that processes textures or mesh data to create a new texture, transferring details from one mesh to another.

Baker Button: A feature in certain 3D modeling software (such as Substance) that allows the user to add a baking operation. Baking in this context means pre-computing and saving certain attributes (like textures) to improve performance.

Baking: The process of transferring details from a high-resolution 3D model to a low-resolution model using texture maps, such as normal maps or ambient occlusion maps.

Base Color: The primary color of a surface, used as a starting point for texture creation.

Beauty Shots: Carefully composed images meant to showcase the best aspects of a scene or object in a portfolio.

Beech Tree: A type of deciduous tree used as an example in this chapter for procedural modeling.

Bezier Curve: A type of curve used in graphic design and 3D modeling to create smooth and scalable lines, defined by a set of control points.

Blend If: A feature in Photoshop that allows blending of layers based on pixel luminosity.

Blend Mode: A setting in image editing software that determines how a layer blends with the layers beneath it, affecting the final appearance based on different mathematical formulas.

Blend Node: A node in Substance Designer used to combine two or more inputs.

Blender: An open-source 3D creation suite that supports the entire 3D pipeline, including modeling, rigging, animation, simulation, rendering, compositing, and motion tracking.

Blockout: An initial, simple version of a 3D model used to define the basic shapes and proportions.

Blur HQ Grayscale Node: A node in Substance Designer used to blur grayscale images.

Boolean Tools: Tools in 3D modeling software used to perform operations like union, subtraction, and intersection on two or more meshes.

Bounding Box: A rectangular box that completely encloses a 3D model, used for calculating the model's spatial dimensions and positioning within a scene.

Branch Generator: A tool in SpeedTree that creates branches according to procedural rules, used to add branches to the tree in different segments like the crown and body.

Burl: A knot or bump in tree trunks, used to add visual interest to long, straight shapes.

Canopy: The upper layer of foliage in a tree or forest.

Channel: In texture mapping, a channel refers to the different components of a texture that can hold different types of data, such as red, green, blue, and alpha (opacity).

Cine Camera Actor: A camera object in Unreal Engine that simulates a real-world camera, allowing for cinematic shots and angles.

Clamp Node: A node used to restrict values within a specified range in a material graph, ensuring that output values do not exceed certain limits.

Clone Patch Node: A node that allows users to clone parts of a texture and use them to cover up unwanted areas, helping to eliminate repetitive elements in a tiling texture.

Color Equalizer Node: A node in software like Substance Designer that adjusts the balance of colors in a texture to create a more uniform appearance.

Color Map: A texture that defines the color information of a 3D model, often used as a base layer for further texturing work.

Color Match Node: A node used to match the colors of one texture to another to ensure consistency across different scanned elements.

Components: Individual sections of a landscape in Unreal Engine, which together form the terrain. The number of components affects the overall size and detail of the terrain.

Composition: The arrangement of elements in a scene to create a visually pleasing or meaningful image.

Content Browser: A window in Unreal Engine that allows users to organize, view, and manage their assets, such as models, textures, and materials, within a project.

Crepuscular Rays (God Rays): Beams of sunlight that appear to radiate from a single point in the sky, often seen in scenes with fog or haze.

Ctrl + A: A shortcut in Blender used to apply transformations to selected objects, such as location, rotation, and scale.

Ctrl + L: A shortcut in Blender for linking data between selected objects, such as modifiers, materials, or animations.

Ctrl + T: A shortcut in Blender for adjusting the tilt or twist of selected curve points.

Curve Modifier: A tool in 3D modeling software that allows a mesh to be deformed according to a curve, making it possible to create complex shapes more easily.

Cutouts/Meshes: Options in SpeedTree for managing and linking meshes and materials.

Data Tab: A panel in 3D modeling software where users can adjust data-related settings of a mesh, such as UV maps and vertex groups.

Decimate Modifier: A tool in Blender that reduces the number of polygons in a mesh while preserving its overall shape, used to create lower-level detail (LOD) models.

Decimation: The process of reducing the number of polygons in a 3D model to make it more efficient for rendering, often used for creating LODs.

Delighting: The process of removing baked-in lighting and shadow information from a texture to make it more neutral and adaptable to different lighting conditions in a game engine.

Density: In the context of Unreal Engine's Foliage Painter, it refers to the number of instances of an asset (like trees) painted in a specific area.

Dilation Width: A parameter that controls the width of the border of a texture, often used to avoid visual seams when textures are applied to models.

DirectionalLight: A type of light in Unreal Engine that simulates sunlight, with parameters for intensity and angle.

Distance Node: A node in Substance Designer used to generate distance-based effects.

Displacement: A method of adding detail to surfaces by modifying the mesh based on texture data.

Divide Node: A node used to perform division operations in a material graph, often employed to scale or normalize values.

Dot Product: A mathematical operation used to calculate the cosine of the angle between two vectors, often employed to determine the similarity or alignment of vectors in a material graph.

DSLR (Digital Single-Lens Reflex Camera): A digital camera combining the optics and mechanisms of a single-lens reflex camera with a digital imaging sensor, often used in photogrammetry for capturing high-resolution images.

ExponentialHeightFog: A tool in Unreal Engine that adds fog to a scene, with settings for density and color, used for creating depth and atmosphere.

Extend Parent Type: A setting in SpeedTree that defines how new branches are generated from existing ones.

Foliage Painter: A tool in Unreal Engine used to paint vegetation and other assets onto a landscape or meshes.

FoliageType: A specific setting in Unreal Engine that contains the parameters for foliage, including density, scaling, and alignment.

Foreground (Fore): The area in an image closest to the viewer, often used to showcase finer details.

Freehand Mode: A mode in SpeedTree that enables drawing and modifying tree elements manually.

Gen Tab: A tab in SpeedTree that contains settings related to the generation of tree branches and elements.

Geometry: The 3D shape and structure of an object, defined by vertices, edges, and faces.

Golden Path: The main route or series of elements in a project that a viewer or player is most likely to focus on.

Grass Clusters: Groups of grass blades arranged together.

Grass Strands: Individual blades of grass.

Ground Cover: The layer of vegetation covering the forest floor.

Grounding Assets: The technique of placing assets in a scene in a way that makes them appear naturally integrated and connected to their environment.

High-Resolution Screenshot: A tool in Unreal Engine for capturing images at higher than screen resolution for enhanced detail.

Histogram: A graphical representation of the tonal values of an image, showing the distribution of brightness levels from dark to light, used for analyzing and adjusting exposure and contrast.

Histogram Scan: The process of capturing detailed information about an object or scene, often using photogrammetry or laser scanning to create accurate 3D models.

Invert Grayscale Node: A node that inverts the grayscale values of an image, swapping dark and light areas, often used to reverse the direction of perceived depth.

Landscape Material: A type of material in Unreal Engine used specifically for terrains. It can automatically apply textures based on various factors like elevation and slope angle.

Layer Mask: A tool in image editing software that allows for non-destructive editing by controlling the transparency of different parts of a layer, revealing or hiding underlying layers.

Levels Node: A node in Substance Designer used to adjust the brightness and contrast of an image.

Lighting: The use of light sources to create depth, focus, and mood within an image or composition.

Live Unwrap: A feature in Blender used to unwrap UVs in real-time.

Low-Poly Mesh: A simplified 3D model with fewer polygons, optimized for performance in real-time applications like video games.

Make It Tile Node: A node used to make textures tile seamlessly across a surface.

Manual Approach: A traditional method of creating content by hand, without automation.

Master Material: A base material in a material system that can be customized and instantiated to create variations.

Material Editor: A tool or interface in 3D software used to create and modify materials by adjusting properties and connecting nodes.

Material Instance: A specific version of a material that allows adjustments to certain parameters without affecting the original material.

Megascans: A library of scanned 3D assets, surfaces, and vegetation used for realistic texturing in Unreal Engine.

Mesh Detail: The level of complexity in a 3D model, defined by the number of polygons and vertices.

Mesh Simplification: The process of reducing the complexity of a 3D model to improve performance or create a lower-resolution version.

Midground (Mid): The section of an image between the foreground and background, usually containing the main subject of focus.

Node-Based Workflow: A method of working in software where different functions are represented as nodes and connected in a graph to create complex effects or materials.

Normal Map: A texture map used to simulate surface detail and texture by altering the way light interacts with the surface, creating the appearance of bumps and grooves without changing the underlying geometry.

Normal Map Node: A node in 3D software used to apply normal maps to a material or surface.

Normal Distribution: The variation of surface normals in a model, affecting how light interacts with the surface.

Object Information Node: A node in Blender that provides information about the selected object, such as its dimensions or material properties.

Opacity: The measure of how transparent or opaque a surface is, defined by an alpha channel in a texture.

Opacity Map: A texture map that defines the transparency levels of different parts of a 3D model's surface.

OpencolorIO (OCIO): A color management system that provides a consistent color experience across different devices and software applications.

Output Node: A node that determines how the final result of a material or texture graph is displayed or rendered.

Paint: A tool or process used to apply color or texture details to a 3D model or 2D image.

Paint Layer: A layer in an image editing or 3D painting application where users can add and modify texture details.

Parallax Effect: A visual phenomenon where objects at different distances appear to move at different speeds when the camera moves.

Parallax Occlusion Mapping: A technique that simulates depth and surface detail by displacing texture coordinates based on view angle, creating a more realistic appearance of surface relief.

Perspective View: A view mode in 3D software that simulates the way objects appear in real life, with parallel lines converging toward a vanishing point.

Photogrammetry: The process of capturing real-world objects and scenes using photos, converting them into 3D models.

Pixel Art: Artwork created at a low resolution and scaled up, where individual pixels are used to define shapes and colors, often associated with retro video games.

Point Lights: A specific type of lighting in 3D composition used to illuminate a particular area or object without casting shadows.

Post-Processing Volume: A tool in Unreal Engine used to adjust visual effects across a scene, such as exposure, saturation, and contrast.

Procedural Generation: A method of creating assets or environments algorithmically, rather than manually, allowing for diverse and complex results based on input parameters.

Projected Texture Mapping: A technique that projects a texture onto a 3D model from a specific viewpoint, similar to how a projector casts an image onto a surface.

Quixel Bridge: A plugin integrated into Unreal Engine that provides access to the Megascans library, a vast collection of high-quality, photorealistic assets and materials.

Random Pitch Angle: A setting that allows slight variations in the angle of assets to make them look more natural and less uniformly aligned.

Remesh: The process of reconstructing the geometry of a 3D model to improve its topology or to create a new mesh with different resolution.

Rim Light: A light used to highlight the edges of an object to separate it from the background.

Root Generator: A tool or feature in SpeedTree that creates and positions root structures for a tree model.

Roughness Map: A texture map that defines the roughness or smoothness of a surface, affecting how light reflects off it and influencing the material's appearance.

Scale Node: A node used to adjust the scale of input values or textures in a material graph.

Scale X (Uniform Scaling): A parameter in Unreal Engine that controls the size of an asset along the X-axis, ensuring proportional scaling when applied uniformly.

Sculpting: The process of shaping and detailing a 3D model using digital tools, similar to working with clay in traditional sculpting.

Sequencer: A tool in Unreal Engine used to create and control cinematic sequences, including animations and camera movements.

Shader: A program that defines how a surface or material is rendered, including how it interacts with light, shadows, and other environmental factors.

Shader Graph: A visual programming interface used to create shaders by connecting nodes that define different aspects of a material's appearance and behavior.

SkyLight: Provides ambient light in a scene and influences the color and intensity of shadows.

Smart Material: A pre-configured material in software like Substance Painter that includes multiple layers and effects, designed to be easily customized for different surfaces.

SpeedTree: A software tool used for creating and modeling trees and vegetation, commonly used in game development and visual effects.

Spot Healing Brush: A Photoshop tool used to remove imperfections or unwanted artifacts in an image.

Substance Designer: A tool used for creating and authoring procedural textures and materials through a node-based interface.

Technical Debt: Unresolved issues or suboptimal solutions put off for later correction during the development process.

Texture Atlas: A single texture that contains multiple images or texture maps, used to reduce the number of texture files and draw calls in a 3D scene.

Texture Baking: The process of creating texture maps that store details like lighting, shadows, and surface detail from a high-resolution model and applying them to a lower-resolution model.

Texture Coordinate Mapping: The process of mapping a 2D texture onto a 3D surface, ensuring that the texture aligns correctly with the geometry.

Texture Painting: The process of manually painting textures directly onto a 3D model, allowing for detailed and custom surface details.

Tiling: The repetition of a texture across a surface, where the texture is duplicated to cover large areas.

Topology: The arrangement and connectivity of vertices, edges, and faces in a 3D model, affecting how the model deforms and interacts with other elements.

UV Mapping: The process of unwrapping a 3D model's surface into a 2D layout, allowing for the application of textures across the model.

Vertex: A point in 3D space that defines the corners of polygons in a mesh, contributing to the overall shape and structure of a 3D model.

Viewport: The area in 3D software where the user can view and interact with their scene or model from different angles and perspectives.

Visual Container: A technique in composition where elements are arranged to create a defined visual space, guiding the viewer's focus and containing the composition.

Volumetric Fog: A more realistic type of fog in Unreal Engine that reacts to light and creates effects like crepuscular rays or "God Rays."

Volumetric Scattering Intensity: A setting in the DirectionalLight that determines how light interacts with volumetric fog, affecting the visibility and strength of "God Rays."

Weight Painting: A technique used to assign influence weights to vertices in a 3D model, controlling how bones or deformers affect the mesh during animation.

World Space: The coordinate system used to define the position and orientation of objects within a 3D scene or environment.

Z-Brush: A digital sculpting tool used for creating highly detailed 3D models, often used for character design and other complex shapes.

Zoom Node: A node used to adjust the zoom level of an input texture or effect in a material graph.

Index

Note: *Italic* page numbers refer to figures.